William Woolsey Johnson

A Treatise on Ordinary and Partial Differential Equations

Third Edition

William Woolsey Johnson

A Treatise on Ordinary and Partial Differential Equations
Third Edition

ISBN/EAN: 9783337811396

Printed in Europe, USA, Canada, Australia, Japan

Cover: Foto ©berggeist007 / pixelio.de

More available books at **www.hansebooks.com**

A TREATISE

ON

ORDINARY AND PARTIAL

DIFFERENTIAL EQUATIONS

BY

WILLIAM WOOLSEY JOHNSON

Professor of Mathematics at the United States Naval Academy
Annapolis Maryland

THIRD EDITION

Second Thousand.

NEW YORK:
JOHN WILEY & SONS,
53 East Tenth Street.
1893.

Entered according to Act of Congress, in the year 1889, by
WILLIAM WOOLSEY JOHNSON,
in the Office of the Librarian of Congress, at Washington.

PREFACE.

THE treatment of the subject of Differential Equations here presented will, it is hoped, be found complete in all those portions which bear upon their practical applications, and in the discussion of their theory so far as it can be adequately treated without the use of the complex variable. The topics included and the order pursued are sufficiently indicated by the table of contents.

An amount of space somewhat greater than usual has been devoted to the geometrical illustrations which arise when the variables are regarded as the rectangular coordinates of a point. This has been done in the belief that the conceptions peculiar to the subject are more readily grasped when embodied in their geometric representations. In this connection the subject of singular solutions of ordinary differential equations and the conception of the characteristic in partial differential equations may be particularly mentioned.

Particular attention has been paid to the development of symbolic methods, especially in connection with the operator $x\dfrac{d}{dx}$, for which, in accordance with recent usage, the symbol ϑ has been adopted. Some new applications of this symbol have been made.

The expression "binomial equations" is applied in this work (in a sense introduced by Boole) to those linear equations which are included in the general form $f_1(\vartheta)y + x^r f_2(\vartheta)y = 0$, and which constitute the class of equations best adapted to solution by development in series. In the sections treating of this method a uniform process has been adopted for the secondary or logarithmic solutions which occur in certain cases. The development of the particular integral when the second member is a power of x is also considered. Chapter VIII is devoted to the general solution of the binomial equation in the notation of the hypergeometric series, and Chapter IX to Riccati's, Bessel's and Legendre's equations.

The examples at the ends of the sections have been derived from various sources, and not a few prepared expressly for this work. They are arranged in order of difficulty, and the solutions are given. These have been verified in the proof-sheets, so that it is believed that they will be found free from errors.

The ordinary references in the text are to Rice and Johnson's Diff. Calc. and Johnson's Int. Calc., published by John Wiley and Sons uniformly with the present volume.

W. W. J.

U. S. NAVAL ACADEMY,
   May, 1889.

CONTENTS.

CHAPTER I.

NATURE AND MEANING OF A DIFFERENTIAL EQUATION BETWEEN TWO VARIABLES.

I.

	PAGE
Solutions in the Form $y = F(x)$	1
Solutions not in the Form $y = F(x)$	3
Particular and Complete Integrals	3
Primitive of a Differential Equation	5
Number of Arbitrary Constants	6
Geometrical Illustration of the Meaning of a Differential Equation	8
Systems of Curves containing an Arbitrary Parameter	9
Doubly Infinite Systems of Curves	10
Examples I.	12

CHAPTER II.

EQUATIONS OF THE FIRST ORDER AND DEGREE.

II.

Separation of the Variables	14
Reduction of the Integral to Algebraic Form	15
Homogeneous Equations	16
Similar and Similarly situated Systems of Curves	18
Case in which the Coefficients of dx and dy are of the First Degree	19
Examples II.	20

III.

Exact Differential Equations	22
Integrating Factors	24

	PAGE
Expressions of the Form $x^\alpha y^\beta(mydx + nxdy)$	27
Examples III	29

IV.

The Linear Equation of the First Order	30
Transformation of a Differential Equation	32
Extension of the Linear Equation	33
Examples IV	34

CHAPTER III.

EQUATIONS OF THE FIRST ORDER, BUT NOT OF THE FIRST DEGREE.

V.

Decomposable Equations	37
Equations Properly of the Second Degree	38
Systems of Curves corresponding to Equations of Different Degrees	39
Standard Form of the Integral of an Equation of the Second Degree	42
Singular Solutions	43
The Discriminant	45
Cusp-Loci	46
Tac-Loci and Node-Loci	47
Examples V	50

VI.

Solution by Differentiation	52
Equations from which One of the Variables is Absent	54
Homogeneous Equations not of the First Degree in p	57
The Equation of the First Degree in x and y	58
Clairaut's Equation	59
Examples VI	61

VII.

Geometrical Applications	63
Polar Coordinates	64
The required Curve a Singular Solution	66
Orthogonal Trajectories	67
Examples VII	69

CHAPTER IV.

EQUATIONS OF THE SECOND ORDER.

VIII.

	PAGE
Successive Integration	72
The First Integrals	74
Integrating Factors	76
Singular Solutions of Equations of the Second Order (foot-note)	77
Derivation of the Complete Integral from Two First Integrals	78
Exact Equations of the Second Order	80
Equations in which y does not occur	82
Equations in which x does not occur	83
The Method of Variation of Parameters	84
Examples VIII.	87

CHAPTER V.

LINEAR EQUATIONS WITH CONSTANT COEFFICIENTS.

IX.

Properties of the Linear Equation	91
The Linear Equation with Constant Coefficients and Second Member Zero	93
Case of Equal Roots	95
Case of Imaginary Roots	97
The Linear Equation with Constant Coefficients and Second Member a Function of x	98
The Inverse Operative Symbol	99
General Expression for the Integral	101
Examples IX.	103

X.

Symbolic Methods of Integration	106
The Second Member X of the Form e^{ax}	106
Case in which X contains a Term of the Form $\sin ax$ or $\cos ax$	109
Case in which X contains Terms of the Form x^m	112
Symbolic Formula of Reduction for the Form $e^{ax}V$	114
Application to the Evaluation of an Ordinary Integral (see also p. 118)	116
Symbolic Formula of Reduction for the Form xV	116
Symbolic Formula of Reduction for the Form $x^r V$	118
Employment of the Exponential Forms of $\sin ax$ and $\cos ax$	119
Examples X.	120

CHAPTER VI

LINEAR EQUATIONS WITH VARIABLE COEFFICIENTS.

XI.

	PAGE
The Homogeneous Linear Equation	124
The Operative Symbol ϑ	125
Complete Integral of the Equation $f(\vartheta)y = 0$	127
Cases of Equal and Imaginary Roots	127
The Particular Integral of $f(\vartheta)y = X$	128
Case in which X is of the Form x^a	129
Symbol Solutions of Linear Equations with Variable Coefficients	130
Non-Commutative Symbolic Factors	132
Examples XI	134

XII.

Exact Linear Equations	135
The Condition of Direct Integrability	137
Integrating Factors of the Form x^m	139
Symbolic Treatment of Exact Linear Equation	140
Symbolic Formulae involving D and ϑ	141
Examples XII	144

XIII.

The Linear Equation of the Second Order	147
Case in which an Integral y_1, when the Second Member is Zero, is known	147
Expression for the Complete Integral in Terms of y_1	149
Relation between Two Independent Integrals y_1 and y_2	151
Symmetrical Expression for the Particular Integral	152
Resolution of the Operator into Factors	153
The Related Equation of the First Order	154
The Transformation $y = vf(x)$	155
The Transformation $y = e^{ax^m}v$	157
Removal of the Term containing the First Derivative — the Normal Form	158
The Invariant for the Transformation $y = vf(x)$ (see also Ex. 26, p. 165)	159
Change of the Independent Variable	160
Examples XIII	162

CHAPTER VII.

SOLUTIONS IN SERIES.

XIV.
                                                                        PAGE
Development of the Integral of a Differential Equation in Series............ 166
Development of the Independent Integrals of a Linear Equation whose Second
    Member is Zero.. 167
Convergency of the Series.. 171
Development of the Particular Integral 172
Binomial and Polynomial Equations 173
Finite Solutions 174
    Examples XIV. 177

XV.

Development of the Logarithmic Form of the Second Integral — Case of Equal
    Values of m.. 181
Case in which the Values of m differ by a Multiple of s................... 185
Special Forms of the Particular Integral 191
    Examples XV. 194

CHAPTER VIII.

THE HYPERGEOMETRIC SERIES.

XVI.

General Solution of the Binomial Equation of the Second Order............ 198
Differential Equation of the Hypergeometric Series........................ 201
Integral Values of γ and γ'.. 202
The Supplementary Series when y_1 is a Finite Series...................... 204
Imaginary Values of α and β.. 205
Infinite Values of α and β.. 206
Case in which α or β equals γ or Unity................................... 208
The Binomial Equation of the Third Order................................ 209
Development of the Solution in Descending Series 210
Transformation of the Equation of the Hypergeometric Series.............. 211
Change of the Independent Variable...................................... 214
The Twenty-Four Integrals... 216
Solutions in Finite Form... 218
    Examples XVI. 220

CHAPTER IX.

SPECIAL FORMS OF DIFFERENTIAL EQUATIONS.

XVII.

	PAGE
Riccati's Equation	224
Standard Linear Form of the Equation	225
Finite Solutions	228
Relations between the Six Integrals	230
Transformations of Riccati's Equation	232
Bessel's Equation	234
Finite Solutions	235
The Besselian Functions	237
The Besselian Functions of the Second Kind	239
Legendre's Equation	241
The Legendrean Coefficients	243
The Second Integral Q_n when n is an Integer	244
Examples XVII.	247

CHAPTER X.

EQUATIONS INVOLVING MORE THAN TWO VARIABLES.

XVIII.

Determinate Systems of the First Order	251
Transformation of Variables	252
Exact Equations	253
The Integrals of a System	254
Equations of Higher Orders equivalent to Determinate Systems of the First Order	256
Geometrical Meaning of a System involving Three Variables	257
Examples XVIII.	258

XIX.

Simultaneous Linear Equations	260
Number of Arbitrary Constants	263
Introduction of a New Variable	264
Examples XIX.	266

XX.

	PAGE
Single Differential Equations involving more than Two Variables	270
The Condition of Integrability	270
Solution of the Integrable Equation	272
Separation of the Variables	274
Homogeneous Equations	275
Equations containing more than Three Variables	276
The Non-Integrable Equation	278
Monge's Solution	280
Geometrical Meaning of a Single Differential Equation between Three Variables	281
The Auxiliary System of Lines	282
Distinction between the Two Cases	283
Examples XX	284

CHAPTER XI.

PARTIAL DIFFERENTIAL EQUATIONS OF THE FIRST ORDER.

XXI.

Equations involving a Single Partial Derivative	287
Equations of the First Order and Degree	288
Lagrange's Solution	289
Geometrical Illustration of Lagrange's Solution	291
Orthogonal Surfaces	292
The Complete and General Primitives	293
Derivation of the Differential Equation from the General Primitive	294
Examples XXI	297

XXII.

The Non-Linear Equation of the First Order	299
The System of Characteristics	300
The General Integral	304
Derivation of a Complete Integral from the Equations of the Characteristic	306
Relation of the General to the Complete Integral	309
Singular Solutions	311
Integrals having Single Points of Contact with the Singular Solution	312
Derivation of the Singular Solution from the Differential Equation	313
Equations involving p and q only	314
Integrals formed by Characteristics passing through a Common Point	314
Equation Analogous to Clairaut's	315
Equations not containing x or y	316

Equations of the Form $f_1(x, p) = f_2(y, q)$ 317
Change of Form in the Equations of the Characteristic 319
Transformation of the Variables 321
 Examples XXII 323

CHAPTER XII.

PARTIAL DIFFERENTIAL EQUATIONS OF HIGHER ORDER.

XXIII.

Equations of the Second Order 326
The Primitive containing Two Arbitrary Functions 326
Forms which give Rise to Equations of the Second Order 327
The Intermediate Equation of the First Order 329
Successive Integration 330
Monge's Method 332
Integrability of Monge's Equations 335
Illustrative Examples 335
 Examples XXIII 340

XXIV.

Linear Equations 341
Homogeneous Equations with Constant Coefficients 342
Symbolic Solution of the Component Equations of the Form $(D - mD')z = 0$ .. 344
Case of Equal Roots 345
Case of Imaginary Roots 347
The Particular Integral 348
The Second Member of the Form $\phi(ax + by)$ 350
The Non-Homogeneous Equation 352
Special Forms of the Integral 354
Special Methods for the Particular Integral 355
The Second Member of the Form $e^{ax + by}$ 356
The Second Member of the Form $\sin(ax + by)$ or $\cos(ax + by)$ 357
The Second Member of the Form $x^r y^s$ 358
The Second Member of the Form $e^{ax + by} V$ 360
Linear Equations with Variable Coefficients 361
The Equation $F(\vartheta, \vartheta')z = 0$ 363
The Equation $F(\vartheta, \vartheta')z = V$ 364
The Symbol $\vartheta + \vartheta'$ in Relation to the Homogeneous Function of x and y 365
 Examples XXIV 366

DIFFERENTIAL EQUATIONS.

CHAPTER I.

NATURE AND MEANING OF A DIFFERENTIAL EQUATION BETWEEN TWO VARIABLES.

I.

Solutions in the Form $y = F(x)$.

1. IN the Integral Calculus, we suppose the differential of a variable y to be given in terms of another variable x and its differential dx, and we seek to express y as a function of x; in other words, since we know that the form of the given equation must be

$$dy = f(x)dx, \quad \dots \dots \dots \quad (1)$$

which may be written

$$\frac{dy}{dx} = f(x), \quad \dots \dots \dots \quad (2)$$

the derivative of y is given in terms of x, and an equation of the form

$$y = F(x) \quad \dots \dots \dots \quad (3)$$

is said to satisfy the given equation (1) or (2) when $F(x)$ is a function whose derivative is the given function $f(x)$.

2. A *differential equation* between two variables x and y is an equation involving in any manner one or more of the derivatives of the unknown function y with respect to x, together with one or both of the variables x and y. The *order* of the equation is that of the highest derivative contained in it, and its *degree* is that of the highest power of this derivative which occurs. An equation of the form $y = F(x)$ satisfies the differential equation if the substitution of $F(x)$ and its derivatives for y and its derivatives reduces it to an identity. For example, the differential equation

$$\frac{d^2y}{dx^2} - 2\frac{dy}{dx} + 2y = 0$$

will be found on trial to be satisfied by $y = e^x \sin x$; for, if we substitute this value for y, and for its derivatives the resulting values $\frac{dy}{dx} = e^x(\cos x + \sin x)$ and $\frac{d^2y}{dx^2} = 2e^x \cos x$, the first member reduces to zero.

3. Equation (1) is, in fact, the simplest form of differential equation. Its general solution is expressed by the formula

$$y = \int f(x)dx; \quad \ldots \ldots \quad (4)$$

and it is the province of the Integral Calculus to reduce this expression, when possible, to a form free from the integral sign, and involving only known functional symbols. But, when this is not possible, the second member of equation (4) represents a new functional form, which, by definition, satisfies equation (1); so that the formula is still the solution of the differential equation. In like manner, a differential equation of any other form is said to be solved when a proper expression is found, even though it involve integrals which we are unable to reduce.

Solutions not in the Form $y = F(x)$.

4. A relation between x and y not in the form $y = f(x)$ may satisfy a differential equation. When this is the case, the values of the derivatives employed in verifying will be expressed in terms of x and y; and, when these are substituted in the differential equation, the result is a relation between x and y which should be true in virtue of the integral equation. For example, in order to show that the equation

$$x\left(\frac{dy}{dx}\right)^2 - y\frac{dy}{dx} + a = 0 \quad \ldots \ldots \quad (1)$$

is satisfied by

$$y^2 = 4ax, \quad \ldots \ldots \ldots \quad (2)$$

we differentiate equation (2); thus,

$$y\frac{dy}{dx} = 2a; \quad \ldots \ldots \ldots \quad (3)$$

and, substituting the value of $\frac{dy}{dx}$ from (3) in equation (1), we have

$$x\frac{4a^2}{y^2} - 2a + a = 0.$$

This equation is true by virtue of the integral relation (2); equation (2) is therefore a solution of the given differential equation (1).

Particular and Complete Integrals.

5. If $F(x)$ is a particular value of the integral in equation (4), Art. 3, then

$$y = F(x) + C,$$

where C is a constant to which any value may be assigned, is the general or complete solution of equation (1). Thus, the general solution involves an arbitrary constant which is called the constant of integration. In like manner, a solution of any differential equation is called a *particular integral* of the equation; but the most general solution, which is called the *complete integral*, contains one or more arbitrary *constants of integration*, the manner in which these constants enter the equation depending on the form of the differential equation.

For example, it was noticed in Art. 2 that the differential equation

$$\frac{d^2y}{dx^2} - 2\frac{dy}{dx} + 2y = 0 \quad \ldots \ldots \quad (1)$$

is satisfied by

$$y = e^x \sin x, \quad \ldots \ldots \ldots \quad (2)$$

which is, therefore, a particular integral. It is not difficult, in this case, to infer from this solution the complete integral; for, in the first place, it is evident that, if we multiply the value of y given in equation (2) by the constant C, the values of $\frac{dy}{dx}$ and $\frac{d^2y}{dx^2}$ will also be multiplied by C, so that the result of substitution in the first member of equation (1) will be C times the previous result, and therefore still equal to zero. Thus,

$$y = Ce^x \sin x \quad \ldots \ldots \ldots \quad (3)$$

is a more general solution of the differential equation. Again, since x does not explicitly enter equation (1), and $x + a$, where a is a constant, has the same differential as x,

$$y = Ce^{x+a} \sin(x + a) \quad \ldots \ldots \quad (4)$$

satisfies the equation, and forms a still more general solution.

Expanding $\sin(x + a)$, and putting

$$Ce^a \cos a = A, \quad Ce^a \sin a = B,$$

we may write the solution in the form

$$y = Ae^x \sin x + Be^x \cos x, \quad \ldots \quad (5)$$

in which A and B are two independent arbitrary constants, because C and a are independently arbitrary. We shall see presently that this equation containing two arbitrary constants is the complete integral of equation (1). The particular integral (2) is the result of putting $A = 1$ and $B = 0$ in the complete integral.

Primitive of a Differential Equation.

6. The general solution found in the preceding article may, of course, be verified by the substitution of the values of y, $\dfrac{dy}{dx}$, and $\dfrac{d^2y}{dx^2}$ in the differential equation. Thus, from

$$y = Ae^x \sin x + Be^x \cos x, \quad \ldots \ldots \ldots \quad (1)$$

we get

$$\frac{dy}{dx} = Ae^x(\sin x + \cos x) + Be^x(\cos x - \sin x), \quad . \quad (2)$$

and

$$\frac{d^2y}{dx^2} = 2Ae^x \cos x - 2Be^x \sin x; \quad \ldots \ldots \ldots \quad (3)$$

and, if these values are substituted in the first member of

$$\frac{d^2y}{dx^2} - 2\frac{dy}{dx} + 2y = 0, \quad \ldots \ldots \quad (4)$$

the coefficients of A and B separately reduce to zero, and the equation is satisfied independently of the values of A and B.

It thus appears that the differential equation (4) is the same as the result of eliminating A and B from equations (1), (2), and (3). Equation (1) is, in this point of view, said to be the *primitive* of equation (4).

7. So also any equation containing arbitrary constants is the primitive of a certain differential equation free from those constants.

For example, if, in the equation

$$c^2x - cy + a = 0, \quad \ldots \ldots \ldots (1)$$

c is regarded as an arbitrary constant, we have, by differentiation,

$$c^2 - c\frac{dy}{dx} = 0, \text{ or } \frac{dy}{dx} = c; \quad \ldots \ldots (2)$$

whence, eliminating c, we obtain

$$x\left(\frac{dy}{dx}\right)^2 - y\frac{dy}{dx} + a = 0 \quad \ldots \ldots (3)$$

as the equation of which equation (1) is the primitive.

Again, equation (2), from which a has disappeared by differentiation, is itself the equation derived from equation (1) as a primitive, when a is regarded as an arbitrary, and c as a fixed, constant. But, if both a and c are arbitrary, differentiating again, we have

$$\frac{d^2y}{dx^2} = 0;$$

and, c having disappeared, this is the equation of the second order of which equation (1) is the primitive.

It is evident that, in every case, the number of differentiations necessary, and therefore the index of the order of the differential equation produced, will be the same as the number of constants to be eliminated.

8. Considering now the differential equation as given, the primitive is an integral equation which satisfies it, the constants eliminated being, in the reverse process of finding the integral, the constants of integration; and it is the most general solution, or complete integral, because no greater number of constants could be eliminated without introducing derivatives higher than the highest which occurs in the given equation. For example, the process given in the preceding article shows that

$$c^2 x - cy + a = 0$$

is the complete integral of

$$x\left(\frac{dy}{dx}\right)^2 - y\frac{dy}{dx} + a = 0.$$

It was shown in Art. 4 that this differential equation is satisfied by $y^2 = 4ax$, which, it will be noticed, is not a particular case of the complete integral. Thus, while the complete integral is the most general solution, it does not, in all cases, include all the solutions.

9. We thus see that the complete integral of a differential equation of the first order should contain one constant of integration, that of an equation of the second order should contain two constants, and so on. It is, of course, to be understood that no two of the constants admit of being replaced by a single one. For example, the constants C and a in the equation

$$y = Ce^{x+a}$$

are equivalent to a single arbitrary constant; for, putting $A = Ce^a$, the equation may be written

$$y = Ae^x,$$

hence it is the complete integral of an equation of the first, not of the second, order.

Geometrical Illustration of the Meaning of a Differential Equation.

10. Let x and y in a differential equation be regarded as the rectangular coordinates of a point in a plane; then the derivative $\dfrac{dy}{dx}$ is the tangent of the inclination to the axis of x of the direction in which the point (x, y) is moving. Putting

$$p = \frac{dy}{dx},$$

a differential equation of the first order is a relation between the variables x, y, and p, of which x and y determine the position of the point, and p the direction of its motion. We may assign to x and y any values we choose, and then determine from the equation one or more values of p. We cannot, therefore, regard the differential equation as satisfied by certain points (that is, by certain associated values of x and y); but it is satisfied by certain associated values of x, y, and p, that is, by a point in any position, provided it is moving in the proper direction.

11. Let us now suppose the point (x, y) to start from any assumed initial position, and to move in the proper direction. We have thus a moving point satisfying the differential equation. As the point moves, the values of x and y vary, so that the value of p derived from the equation will likewise, in general, vary; and we may suppose the direction of the point's motion to vary in such a way that the moving point continues to satisfy the differential equation. The line which the point now describes is, in general, a curve; and the point may evidently move along this curve in either direction, and yet always satisfy the differential equation. The moving point may return to its initial position, thus describing a closed curve; or it may pass to infinity in both directions, describing an infinite branch of a curve.

If, now, we can determine the equation of this curve in the form of an ordinary equation between x and y, the value of $\frac{dy}{dx}$ found by differentiating the equation of the curve will, by hypothesis, be identical, for any values of x and y, with the value of p corresponding to the same values of x and y in the differential equation. The equation of the curve will, therefore, be a solution, or integral, of the differential equation.

12. But, since this integral equation restricts the point to certain positions, the assemblage of which constitutes the curve, it is not the complete solution of the differential equation; for the complete solution ought to represent the moving point satisfying the equation in all its possible positions. If, now, we take for initial point any point not on the curve already determined, and proceed in like manner, we shall determine another curve, whose equation will be another particular solution, or integral, of the differential equation. We thus have an unlimited number of curves forming a *system of curves*, and the complete integral is the general equation of this system.

This general equation must contain, besides x and y, a quantity independent of x and y called the *parameter*, by giving different values to which we obtain the equations of all the particular curves of the system. The arbitrary parameter of the system is, of course, the constant of integration.

13. We may illustrate this by a simple example. Let the differential equation be

$$\frac{dy}{dx} = -\frac{x}{y}. \qquad \qquad (1)$$

Since $\frac{y}{x}$ is the tangent of the inclination to the axis of x of the line joining the point (x, y) to the origin, the equation expresses that the point (x, y) is always moving in a direction perpendicular to the line joining it to the origin. Starting from any initial position, it is clear that the point describes a circle about the origin as centre. The system of curves in this case, therefore, consists of all circles whose centres are at the origin; and the general equation of this system,

$$x^2 + y^2 = C, \quad \ldots \quad \ldots \quad \ldots \quad (2)$$

where C is the parameter, is the complete integral.

Now consider the moving point when in any special position, as, for instance, in the position (3, 2); we find, by substituting these values for x and y in equation (2),

$$C = 13.$$

Hence

$$x^2 + y^2 = 13$$

is the equation of the particular curve in which the point is then moving. If we differentiate this equation, we find a value for $\dfrac{dy}{dx}$ at the point (3, 2) identical with that given for the same point by equation (1).

Doubly Infinite Systems of Curves.

14. In the case of a differential equation of the second order, let

$$\frac{dy}{dx} = p \quad \text{and} \quad \frac{d^2y}{dx^2} = q;$$

then the equation is a relation between x, y, p, and q. It is possible to assign any values we please to x, y, and p, and to determine from the equation a value of q, which, in connection with the assumed values of x, y, and p, will satisfy the equation. This value of q, in connection with the assumed value of p, determines the curvature of the path of the moving point

(x, y). Hence a differential equation of the second order may be regarded as satisfied by a moving point having any assumed position, and moving in any assumed direction, provided only that its path have the proper curvature. Starting from any assumed initial point, and in any assumed initial direction, the point (x, y) may move in such a manner as to satisfy the equation. As it moves the values of x and y will vary; and, since the path has a definite curvature at this point, the value of p will likewise vary. Hence the value of q derived from the differential equation will, in general, also vary; but we may suppose the curvature of the path to vary in such a manner that the moving point continues to satisfy the equation. A curve is thus described whose ordinary equation is a solution of the differential equation, since the simultaneous values of x, y, $\frac{dy}{dx}$, and $\frac{d^2y}{dx^2}$, at every point of it, by hypothesis, satisfy that equation.

15. As before, the complete integral is the general equation of the system of curves which may be generated in the manner explained above; but this system has a greater generality than that which represents a differential equation of the first order. For, in its general equation, it must be possible to assign any assumed simultaneous values to x, y, and p. Substituting the assumed values in the general equation and in the result of its differentiation, we have two equations; and, in order to satisfy them, we must have two arbitrary parameters at our disposal.

The system of curves representing a differential equation of the second order is, therefore, a system containing two parameters, to each of which independently an unlimited number of values may be assigned. Such a system is said to be a *doubly infinite* system of curves.

In like manner, it may be shown that a differential equation of the third order is represented by a triply infinite system of curves, and so on.

Examples I.

1. Form the differential equation of which $y = c \cos x$ is the complete integral.

$$\frac{dy}{dx} + y \tan x = 0.$$

2. Form the equation of which $y = ax^2 + bx$ is the complete integral, a and b being arbitrary.

$$x^2 \frac{d^2y}{dx^2} - 2x \frac{dy}{dx} + 2y = 0.$$

3. Form the equation of which $y^2 - 2cx - c^2 = 0$ is the complete integral.

$$y \left(\frac{dy}{dx}\right)^2 + 2x \frac{dy}{dx} - y = 0.$$

4. Form the equation of which $e^{2y} + 2cxe^y + c^2 = 0$ is the primitive.

$$\left(\frac{dy}{dx}\right)^2 (1 - x^2) + 1 = 0.$$

5. Form the equation of which $y = (x + c)e^{ax}$ is the complete integral.

$$\frac{dy}{dx} = e^{ax} + ay.$$

6. Denoting by θ the inclination to the axis of x of the line joining (x, y) to the origin, and by ϕ the inclination of the point's motion, write the differential equation which expresses that ϕ is the supplement of θ, and show that it represents a system of hyperbolas.

7. With the same notation, write the differential equation which expresses that $\phi = 2\theta$, and show that it represents a system of circles passing through the origin.

8. Form the differential equation of the system of straight lines which touch the circle $x^2 + y^2 = 1$, and show that this circle also satisfies the equation.

9. Find the differential equation of all the circles having their radii equal to a.

$$\left[\left(\frac{dy}{dx}\right)^2 + 1\right]^3 = a^2\left(\frac{d^2y}{dx^2}\right)^2.$$

10. Find the differential equation of all the conics whose axes coincide with the coordinate axes.

$$y\frac{dy}{dx} - x\left(\frac{dy}{dx}\right)^2 - xy\frac{d^2y}{dx^2} = 0.$$

CHAPTER II.

EQUATIONS OF THE FIRST ORDER AND DEGREE.

II.

Separation of the Variables.

16. In an equation of the first order, it is immaterial whether x or y be taken as the independent variable. If the equation is also of the first degree, it is frequently written in the form

$$Mdx + Ndy = 0,$$

in which M and N denote functions of x and y. The simplest case is that in which the equation may be so written that the coefficient of dx is a function of x only, and that of y a function of y only; in other words, the case in which the equation can be written in the form

$$f(x)dx + \phi(y)dy = 0. \quad \ldots \ldots \quad (1)$$

The complete integral is then evidently

$$\int f(x)dx + \int \phi(y)dy = C. \quad \ldots \ldots \quad (2)$$

17. The process of reducing an equation, when possible, to the form (1) is called the *separation of the variables*. For example, in the equation

$$(1 - y)dx + (1 + x)dy = 0, \quad \ldots \ldots (1)$$

the variables are separated by dividing by $(1 - y)(1 + x)$; thus,

$$\frac{dx}{1 + x} + \frac{dy}{1 - y} = 0. \quad \ldots \ldots (2)$$

Hence, integrating,

$$\log(1 + x) - \log(1 - y) = c. \quad \ldots \ldots (3)$$

18. The integral here presents itself in a transcendental form; but it is readily reduced to an algebraic form, for (3) may be written in the form

$$\log \frac{1 + x}{1 - y} = c; \quad \ldots \ldots (4)$$

whence

$$\frac{1 + x}{1 - y} = e^c; \quad \ldots \ldots (5)$$

or, putting C for e^c,

$$1 + x = C(1 - y). \quad \ldots \ldots (6)$$

It is to be noticed that C in equation (6) admits of all values positive and negative, although e^c can only be positive. In fact, equation (4) is defective in notation; for, since the integrals are the logarithms of the numerical values of $1 + x$ and $1 - y$ respectively (see Int. Calc., Art. 10), that equation ought strictly to have been written

$$\log\left(\pm \frac{1 + x}{1 - y}\right) = c,$$

and finally C is put for $\pm e^c$.

19. The complete integral in the above example is readily seen to be the equation of a system of straight lines passing through the point $(-1, 1)$. In general, any assumed simultaneous values of x and y, that is, any assumed position of the moving point, determines a value of C, as in Art. 13. But, for the particular point $(-1, 1)$, the value of C is indeterminate in equation (6); and accordingly we find that p or $\frac{dy}{dx}$ is also indeterminate for this point in equation (1).

It must not, however, be assumed that, whenever p in the differential equation is indeterminate at a particular point, all the lines represented by the complete integral pass through the point in question. For, if the point be a node of the particular integral which passes through it, p will have, at this point, more than one value; and, the differential equation being of the first degree, this can only happen when p takes the indeterminate form. For example, the integral of $xdy + ydx = 0$ is $xy = C$, representing a system of hyperbolas; but the particular integral which passes through the origin is the pair of straight lines $xy = 0$ of which the origin is a node. Accordingly p takes the indeterminate form at the origin.

Homogeneous Equations.

20. The differential equation of the first order and degree, $Mdx + Ndy = 0$, is said to be *homogeneous* when M and N are homogeneous functions of x and y of the same degree. Since the ratio of two homogeneous functions of the same degree is a function of $\frac{y}{x}$, a homogeneous equation may be written in the form

$$\frac{dy}{dx} = f\left(\frac{y}{x}\right). \quad \ldots \ldots \ldots (1)$$

§ II.]   HOMOGENEOUS EQUATIONS.

If, now, we put v for $\dfrac{y}{x}$, so that

$$y = vx, \quad \frac{dy}{dx} = x\frac{dv}{dx} + v,$$

the equation becomes

$$x\frac{dv}{dx} + v = f(v),$$

a differential equation between x and v in which the variables can be separated; thus,

$$\frac{dx}{x} = \frac{dv}{f(v) - v}.$$

21. For example, the equation

$$(x^2 - y^2)\frac{dy}{dx} - 2xy = 0 \quad \ldots \ldots \quad (1)$$

is homogeneous. Putting $y = vx$, we obtain

$$x\frac{dv}{dx} + v = \frac{2v}{1 - v^2};$$

whence

$$x\frac{dv}{dx} = \frac{v + v^3}{1 - v^2},$$

or

$$\frac{dx}{x} = \frac{1 - v^2}{v(1 + v^2)}dv = \frac{dv}{v} - \frac{2v\,dv}{1 + v^2}.$$

Integrating,

$$\log x = \log \frac{v}{1 + v^2} + C;$$

and, replacing v by $\dfrac{y}{x}$,

$$\log\frac{x^2 + y^2}{y} = C,$$

or

$$x^2 + y^2 = cy. \quad \dots \quad \dots \quad \dots \quad (2)$$

22. The geometrical meaning of the homogeneous equation (1) of Art. 20 is that $\dfrac{dy}{dx}$ has the same value for all points at which $\dfrac{y}{x}$ has a given value; that is to say, if we draw a straight line through the origin, the various curves of the system represented have all the same direction at their points of intersection with this straight line. But this is the definition of similar and similarly situated curves, the origin being the centre of similitude. The curves of such a system are simply the different curves which would be constructed to represent the same equation if we took different units of length. Denoting the unit of length by c, the general equation of the system will therefore be of the form

$$f\left(\frac{x}{c}, \frac{y}{c}\right) = 0,$$

where, since c is arbitrary, it is the parameter of the system; in other words, the constant of integration c may be so taken that the complete integral of the homogeneous differential equation will be a homogeneous equation between x, y, and c. In the example given in the preceding article, the system of curves represented consists of all circles touching the axis of x at the origin, and the final equation is so written that all of its terms are of the second degree with respect to x, y, and c.

Case in which the Functions M and N are of the First Degree.

23. The equation $Mdx + Ndy = 0$ can always be solved if M and N are functions of the first degree in x and y; that is, when the equation is of the form

$$\frac{dy}{dx} = \frac{ax + by + c}{a'x + b'y + c'}; \quad \dots \dots (1)$$

for, substitute in this

$$\left.\begin{array}{l} x = \xi + h, \\ y = \eta + k, \end{array}\right\} \quad \dots \dots \dots (2)$$

and we have

$$\frac{d\eta}{d\xi} = \frac{a\xi + b\eta + ah + bk + c}{a'\xi + b'\eta + a'h + b'k + c'}. \quad \dots (3)$$

If, now, we determine h and k by the equations

$$\left.\begin{array}{l} ah + bk + c = 0, \\ a'h + b'k + c' = 0, \end{array}\right\}. \quad \dots \dots (4)$$

equation (3) takes the homogeneous form

$$\frac{d\eta}{d\xi} = \frac{a\xi + b\eta}{a'\xi + b'\eta},$$

from which we can determine the integral relation between ξ and η; and thence, by substitution from (2), the relation between x and y.

24. Equations (4) give impossible values of h and k when a, b, a', and b' are proportional. In this case, putting

$$a' = ma, \quad b' = mb,$$

equation (1) becomes

$$\frac{dy}{dx} = \frac{ax + by + c}{m(ax + by) + c'}.$$

Now put

$$ax + by = z;$$

whence

$$\frac{dy}{dx} = \frac{1}{b}\frac{dz}{dx} - \frac{a}{b}.$$

Making these substitutions, we have

$$\frac{dz}{dx} = a + b\frac{z + c}{mz + c'},$$

an equation in which the variables can be separated.

Examples II.

Solve the following differential equations:—

1. $(1 + x)y\,dx + (1 - y)x\,dy = 0,$ $\log xy = c - x + y.$

2. $\dfrac{dy}{dx} = ay^2 x,$ $ax^2 y + cy + 2 = 0.$

3. $(y^2 + xy^2)dx + (x^2 - yx^2)dy = 0,$ $\log \dfrac{x}{y} - \dfrac{y + x}{xy} = c.$

4. $xy(1 + x^2)dy = (1 + y^2)dx,$ $(1 + x^2)(1 + y^2) = cx^2.$

5. $\dfrac{x\,dx}{1 + y} = \dfrac{y\,dy}{1 + x},$ $3(x^2 - y^2) + 2(x^3 - y^3) = c.$

6. $\dfrac{dy}{dx} + b^2 y^2 = a^2,$ $\dfrac{by + a}{by - a} = ce^{2abx}.$

7. $\dfrac{dy}{dx} = \dfrac{y^2 + 1}{x^2 + 1},$ $y - x = c(1 + xy).$

8. $\sin x \cos y \, dx = \cos x \sin y \, dy,$ $\qquad \cos y = c \cos x.$

9. $ax\dfrac{dy}{dx} + 2y = xy\dfrac{dy}{dx},$ $\qquad x^2 y^a = ce^y.$

10. $\dfrac{dy}{dx} + \dfrac{1 + y + y^2}{1 + x + x^2} = 0,$ $\qquad \dfrac{x + y + 1}{2xy + x + y - 1} = c.$

11. $\dfrac{dy}{dx} + e^x y = e^x y^2,$ $\qquad \log \dfrac{y - 1}{y} = e^x + c.$

12. $x\dfrac{dy}{dx}(1 - y^b) = y(1 + x^a),$ $\qquad \log\dfrac{y}{x} - \dfrac{x^a}{a} - \dfrac{y^b}{b} = c.$

13. $xdy - ydx - \sqrt{(x^2 + y^2)}dx = 0,$ $\qquad x^2 = c^2 + 2cy.$

14. $(8y + 10x)dx + (5y + 7x)dy = 0,$
$\qquad\qquad (y + x)^2(y + 2x)^3 = c.$

15. $(x + y)\dfrac{dy}{dx} + x - y = 0,\quad \tan^{-1}\dfrac{y}{x} + \tfrac{1}{2}\log(x^2 + y^2) = c.$

16. $(xy - x^2)\dfrac{dy}{dx} = y^2,$ $\qquad y = ce^{\frac{y}{x}}.$

17. $x + y\dfrac{dy}{dx} = 2y,$ $\qquad \log(x - y) = c - \dfrac{x}{x - y}.$

18. $(3y - 7x + 7)dx + (7y - 3x + 3)dy = 0,$
$\qquad\qquad (y - x + 1)^2(y + x - 1)^5 = c.$

19. $(x^2 + y^2)dx - 2xydy = 0,$ $\qquad x^2 - y^2 = cx.$

20. $2xydx + (y^2 - 3x^2)dy = 0,$ $\qquad x^2 - y^2 = cy^3.$

21. $y^2 + (xy + x^2)\dfrac{dy}{dx} = 0,$ $\qquad xy^2 = c(x + 2y).$

22. $(x^2 - 3y^2)xdx + (3x^2 - y^2)ydy = 0,$
$\qquad\qquad (x^2 + y^2)^2 = c(y^2 - x^2).$

III.

Exact Differential Equations.

25. An *exact differential* containing two variables is an expression which may arise from the differentiation of a function of x and y. Denoting the function by u, we have

$$du = \frac{\partial u}{\partial x} dx + \frac{\partial u}{\partial y} dy, \quad \ldots \quad (1)$$

where the coefficients of dx and dy are the partial derivatives of u. Thus, the form of an exact differential is $Mdx + Ndy$. But, if M and N are any given functions of x and y, we cannot generally put

$$du = Mdx + Ndy; \quad \ldots \quad (2)$$

for, if

$$M = \frac{\partial u}{\partial x}, \text{ and } N = \frac{\partial u}{\partial y}, \quad \ldots \quad (3)$$

we must have

$$\frac{dM}{dy} = \frac{dN}{dx}, \quad \ldots \quad (4)$$

because each member of this equation is an expression for $\frac{d^2u}{dxdy}$.

Hence equation (4) is a necessary condition of the possibility of equation (2) or equations (3); that is, of the existence of a function whose partial derivatives with respect to x and y are M and N respectively.

26. It is also a sufficient condition; for the most general form of the function whose derivative with respect to x is M, is

$$u = \int Mdx + Y, \quad \ldots \quad (5)$$

where Mdx is integrated as if y were constant, and Y is a quantity independent of x, but which may be a function of y.

§ III.] EXACT DIFFERENTIAL EQUATIONS. 23

Now the only other condition to be satisfied is that the derivative of u with respect to y shall equal N; that is,

$$N = \frac{d}{dy}\int M dx + \frac{dY}{dy},$$

or

$$\frac{dY}{dy} = N - \frac{d}{dy}\int M dx. \quad \ldots \quad (6)$$

Since Y is to be a function of y only, but is otherwise unrestricted, this equation merely requires that the second member should be independent of x. This will be true if its derivative with respect to x vanishes; that is to say, if

$$\frac{dN}{dx} - \frac{dM}{dy} = 0.$$

This equation is identical with equation (4), which is, therefore, a sufficient, as well as a necessary, condition.

27. An equation in which an exact differential is equated to zero is called an *exact differential equation*. Using the notation of the preceding articles, the complete integral of the equation $Mdx + Ndy = 0$ when exact is evidently

$$u = C.$$

The function u is determined by direct integration as indicated in equations (5) and (6). It is to be noticed that dY consists of those terms in Ndy which do not involve x; and evidently the integral of these terms, and also of those containing x only, may be considered separately, and it is only necessary to ascertain whether the terms containing both x and y form an exact differential. For example, in the equation

$$x(x + 2y)dx + (x^2 - y^2)dy = 0,$$

the sum of these terms is $2xy\,dx + x^2\,dy$, which is the differential of x^2y; hence the complete integral is

$$\tfrac{1}{3}x^3 + x^2y - \tfrac{1}{3}y^3 = C,$$

or

$$x^3 + 3x^2y - y^3 = c.$$

28. An expression involving only some function of x and y, and the differential of this function, is obviously an exact differential. Thus, in the equation

$$\frac{x\,dx + y\,dy}{\sqrt{(x^2 + y^2 - 1)}} + \frac{y\,dx - x\,dy}{x^2 + y^2} = 0,$$

the first term is a function of $x^2 + y^2$ and its differential, and the second is a function of $\dfrac{x}{y}$ and its differential. The equation may, in fact, be written

$$\frac{d(x^2 + y^2)}{2\sqrt{(x^2 + y^2 - 1)}} + \frac{d\left(\dfrac{x}{y}\right)}{\left(\dfrac{x}{y}\right)^2 + 1} = 0;$$

hence the integral is

$$\sqrt{(x^2 + y^2 - 1)} + \tan^{-1}\frac{x}{y} = C.$$

Integrating Factors.

29. We have seen in the preceding articles that the complete integral of an exact differential equation appears in the form

$$u = C, \qquad \qquad (1)$$

so that the differential equation results directly from the differentiation of the integral, C disappearing by differentiation.

Now, since the integral of any equation can be put in the form (1) by solving it for C, it follows that, whenever we can solve an equation of the form

$$Mdx + Ndy = 0, \quad \ldots \ldots \quad (2)$$

we can produce an exact differential equation which is equivalent to the given equation, that is to say, which is satisfied by the same simultaneous values of x, y, and p. This new differential equation being of the first order and degree, must then be of the form

$$\mu(Mdx + Ndy) = 0, \quad \ldots \ldots \quad (3)$$

where μ is a factor containing x or y or both, but not containing p.

The factor μ, which converts a given differential equation into an exact differential equation, is called an *integrating factor*.

For example, solving equation (2) of Art. 21 for c, we have

$$c = y + \frac{x^2}{y};$$

whence, differentiating,

$$0 = dy + \frac{2xydx - x^2dy}{y^2} = \frac{2xydx + (y^2 - x^2)dy}{y^2};$$

and, comparing this with equation (1) of Art. 21, we see that $\frac{1}{y^2}$ is an integrating factor of that equation.

30. A differential equation has a variety of integrating factors corresponding to different forms of the complete integral. For example, one integrating factor of equation (1),

Art. 17, is the factor by means of which we separated the variables; namely,

$$\frac{1}{(1+x)(1-y)};$$

and this corresponds to the form (3) of the integral; but, if we differentiate the integral in the form (5), Art. 18, we obtain equation (1) multiplied by the integrating factor

$$\frac{1}{(1-y)^2}.$$

The forms of the integral differ in respect to the constants which they contain. In general, if $u = c$ is an integral giving the integrating factor μ, so that

$$du = \mu(Mdx + Ndy),$$

then

$$f(u) = C$$

where $C = f(c)$ is another form of the integral; and this gives the exact differential equation

$$f'(u)du = 0,$$

or

$$f'(u)\mu(Mdx + Ndy) = 0.$$

Hence $f'(u)\mu$ is also an integrating factor; and, since f denotes an arbitrary function, f' is also arbitrary; thus, the number of integrating factors is unlimited.

31. The form of the given differential equation sometimes suggests an integrating factor. For example, in the equation

$$(y + \log x)dx - xdy = 0,$$

the terms containing both x and y are

$$ydx - xdy.$$

This expression becomes an exact differential when divided either by y^2 or by x^2. The remaining term contains x only; hence $\frac{1}{x^2}$ is an integrating factor. Thus, we write

$$\frac{ydx - xdy}{x^2} + \frac{\log x \, dx}{x^2} = 0;$$

whence, integrating,

$$\frac{-y}{x} - \frac{\log x}{x} - \frac{1}{x} = c,$$

or

$$cx + y + \log x + 1 = 0.$$

32. The expression $ydx - xdy$, which occurs in the preceding article, is a special case of a more general one which should be noticed. For, since

$$d(x^m y^n) = x^{m-1} y^{n-1}(my\,dx + nx\,dy),$$

an expression of the form

$$x^\alpha y^\beta (my\,dx + nx\,dy) \quad \ldots \ldots \quad (1)$$

has the integrating factor

$$x^{m-1-\alpha} y^{n-1-\beta};$$

and since, by Art. 30, the product of this by any function of u, where $u = x^m y^n$, is also an integrating factor, we have the more general expression

$$x^{km-1-\alpha} y^{kn-1-\beta} \quad \ldots \ldots \quad (2)$$

(in which k may have any value) for an integrating factor.

As an illustration, take the equation

$$y(y^3 + 2x^4)dx + x(x^4 - 2y^3)dy = 0.$$

This may be written in the form

$$y^3(ydx - 2xdy) + x^4(2ydx + xdy) = 0,$$

in which each term is of the form (1). In the first term,

$$m = 1, \quad n = -2, \quad \alpha = 0, \quad \beta = 3;$$

and the expression (2) gives, for the integrating factor,

$$x^{k-1}y^{-2k-4};$$

that is to say, any multiplier of this form will convert the first term into an exact differential. In like manner, any expression of the form

$$x^{2k'-5}y^{k'-1}$$

is an integrating factor of the second term. A quantity which is at once of each of these forms will therefore be an integrating factor of the given equation. Equating the exponents of x, and also the exponents of y, in the two expressions, we have

$$k - 1 = 2k' - 5,$$

$$-2k - 4 = k' - 1,$$

from which $k = -2$, and the integrating factor is x^{-3}.

Multiplying the given equation by x^{-3}, we have

$$y^4x^{-3}dx - 2y^3x^{-2}dy + 2xydx + x^2dy = 0;$$

and, integrating,
$$\frac{-y^4}{2x^2} + x^2y = C,$$
or
$$2x^4y - y^4 = cx^2.$$

Examples III.

Solve the following differential equations:—

1. $(x^2 - 4xy - 2y^2)dx + (y^2 - 4xy - 2x^2)dy = 0,$
$x^3 + y^3 - 6xy(x + y) = c.$

2. $\dfrac{dy}{dx} = \dfrac{y}{x}\dfrac{2x - y^2}{3y^2 - x},$ $\qquad xy^3 = x^2y + c.$

3. $(2x - y + 1)dx + (2y - x - 1)dy = 0,$
$x^2 - xy + y^2 + x - y = c.$

4. $x(x^2 + 3y^2)dx + y(y^2 + 3x^2)dy = 0,$
$x^4 + 6x^2y^2 + y^4 = c.$

5. $ydy + xdx + \dfrac{xdy - ydx}{x^2 + y^2} = 0,$ $\qquad \dfrac{x^2 + y^2}{2} + \tan^{-1}\dfrac{y}{x} = c.$

6. $(y - x)dy + ydx = 0,$ $\qquad \log y + \dfrac{x}{y} = c.$

7. $ax^2y^n\dfrac{dy}{dx} = 2x\dfrac{dy}{dx} - y,$ $\qquad \dfrac{ay^{n+2}}{n+2} = \dfrac{y^2}{x} + c.$

8. $x\dfrac{dy}{dx} - y = \sqrt{(x^2 - y^2)},$ $\qquad \sin^{-1}\dfrac{y}{x} = \log x + c.$

9. $x\dfrac{dy}{dx} - y = x\sqrt{(x^2 + y^2)},$ $\qquad y = \dfrac{x}{2}\left(ce^x - \dfrac{1}{ce^x}\right).$

10. $\cos ax \dfrac{dy}{dx} + ay \sin ax = x,$

$\qquad a^2 y = ax \sin ax + \cos ax \cdot \log \cos ax + c \cos ax.$

11. $(y^3 - 2yx^2) dx + (2xy^2 - x^3) dy = 0, \qquad x^2 y^4 = y^2 x^4 + c.$

12. $(2x^2 y^2 + y) dx - (x^3 y - 3x) dy = 0, \qquad 4x^2 y = 5 + cx^{\frac{1}{2}} y^{\frac{1}{2}}.$

13. $(y^4 - 2x^3 y) dx + (x^4 - 2xy^3) dy = 0, \qquad x^3 + y^3 = cxy.$

14. Solve Exs. II., 19 and 20, by means of integrating factors.

IV.

The Linear Equation of the First Order.

33. A differential equation is said to be *linear* when it is of the first degree with respect to y and its derivatives. The linear equation of the first order may therefore be written in the form

$$\dfrac{dy}{dx} + Py = Q,$$

where P and Q denote functions of x.

Consider first the case in which the second member is zero, that is to say, the form

$$\dfrac{dy}{dx} + Py = 0 \quad \dotfill \quad (1)$$

The variables may be separated; thus,

$$\dfrac{dy}{y} = -P dx.$$

§ IV.]   LINEAR EQUATION OF THE FIRST ORDER.   31

Hence
$$\log y = c - \int P dx,$$
or
$$y = Ce^{-\int P dx}. \quad \ldots \ldots \ldots \quad (2)$$

In this solution, $\int P dx$ may be taken to denote a particular integral of $P dx$, since the constant is directly expressed in the equation.

34. If we put equation (2) in the form

$$e^{\int P dx} y = C,$$

and differentiate, we have

$$e^{\int P dx}(dy + Pydx) = 0,$$

which shows that $e^{\int P dx}$ is an integrating factor of equation (1). Since Q is a function of x only, it follows that $e^{\int P dx}$ is also an integrating factor of the more general equation

$$\frac{dy}{dx} + Py = Q. \quad \ldots \ldots \ldots \quad (1)$$

Hence, to solve this equation, we write

$$e^{\int P dx}(dy + Pydx) = e^{\int P dx} Q dx;$$

and, integrating,

$$e^{\int P dx} y = \int e^{\int P dx} Q dx + C. \quad \ldots \ldots \quad (2)$$

In a given example, the integrals involved in the general expression (2) should, of course, be evaluated if possible.

Thus, let the given equation be

$$(1 + x^2)\frac{dy}{dx} = m + xy,$$

or

$$\frac{dy}{dx} - \frac{x}{1 + x^2}y = \frac{m}{1 + x^2}.$$

Here $P = -\dfrac{x}{1 + x^2}$; therefore

$$\int P dx = -\tfrac{1}{2} \log(1 + x^2),$$

and

$$e^{\int P dx} = \frac{1}{\sqrt{(1 + x^2)}}$$

is the integrating factor. Hence

$$\frac{1}{\sqrt{(1 + x^2)}} dy - \frac{x}{(1 + x^2)^{\frac{3}{2}}} y dx = \frac{m dx}{(1 + x^2)^{\frac{3}{2}}},$$

and

$$\frac{y}{\sqrt{(1 + x^2)}} = \frac{mx}{\sqrt{(1 + x^2)}} + C,$$

or

$$y = mx + C\sqrt{(1 + x^2)}.$$

Transformation of a Differential Equation.

35. It is frequently useful to transform a given differential equation by replacing one of the variables by a new variable which is an assumed function of the variable replaced, or of both variables. The form of the assumed function is generally suggested by that of the given equation. Thus, the form of the equation

$$(1 + xy)y\,dx + (1 - xy)x\,dy = 0$$

suggests the use of a new variable $v = xy$, whence

$$x\,dy = dv - y\,dx.$$

Eliminating y, we have

$$(1 + v)\frac{v}{x}dx + (1 - v)dv - (1 - v)\frac{v}{x}dx = 0,$$

or

$$2v^2 dx + (1 - v)x\,dv = 0,$$

in which the variables v and x can be separated. Integrating, we obtain

$$2\log x - \frac{1}{v} - \log v = c,$$

or

$$\frac{x^2}{v} = Ce^{\frac{1}{v}},$$

and, substituting xy for v,

$$x = Cye^{\frac{1}{xy}}.$$

Extension of the Linear Equation.

36. If $v = f(y)$, the linear equation for v,

$$\frac{dv}{dx} + Pv = Q,$$

becomes

$$f'(y)\frac{dy}{dx} + Pf(y) = Q. \quad \cdots \cdots \quad (1)$$

In other words, an equation of this form becomes linear if we put $v = f(y)$.

For example, the equation

$$\frac{dy}{dx} + \frac{\tan y}{x+1} - (x-1)\sec y = 0$$

takes the form (1) when multiplied by $\cos y$, and hence is a linear equation for $\sin y$. The integral will be found to be

$$\sin y = \frac{x^3 - 3x + c}{3(x+1)}.$$

37. In particular, the equation   *Bernoulli's*

$$\frac{dy}{dx} + Py = Qy^n \quad \ldots \ldots \ldots \quad (2)$$

is known as the extension of the linear equation. Dividing by y^n, we have

$$y^{-n}\frac{dy}{dx} + Py^{1-n} = Q,$$

or

$$(1-n)y^{-n}\frac{dy}{dx} + (1-n)Py^{1-n} = (1-n)Q,$$

which is of the form (1), and therefore linear for y^{1-n}.

EXAMPLES IV.

Solve the following linear equations:—

✓ 1. $\dfrac{dy}{dx} + y = x,$ $\qquad y = x - 1 + ce^{-x}.$

2. $\dfrac{dy}{dx} = by + a\sin x,$ $\qquad y = ce^{bx} - a\dfrac{b\sin x + \cos x}{1 + b^2}.$

✓ 3. $\dfrac{dy}{dx} - \dfrac{2y}{x+1} = (x+1)^3,$ $\qquad 2y = (x+1)^4 + c(x+1)^2.$

§ IV.] EXAMPLES. 35

4. $\dfrac{dy}{dx} - n\dfrac{y}{x} = e^x x^n,$ $y = x^n(e^x + c).$

✓ 5. $\dfrac{dy}{dx} - xy = x,$ $y = ce^{\frac{1}{2}x^2} - 1.$

6. $\dfrac{dy}{dx}\cos x + y \sin x = 1,$ $y = \sin x + c \cos x.$

✓ 7. $\dfrac{dy}{dx} + y \cos x = \sin 2x,$ $y = 2 \sin x - 2 + ce^{-\sin x}.$

8. $x\dfrac{dy}{dx} - ay = x + 1,$ $y = \dfrac{x}{1-a} - \dfrac{1}{a} + cx^a.$

✓ 9. $\dfrac{dy}{dx} + \dfrac{xy}{1+x^2} = \dfrac{1}{2x(1+x^2)},$

$$y = \dfrac{\log[\sqrt{(1+x^2)} - 1] - \log x + c}{2\sqrt{(1+x^2)}}.$$

10. $x(1-x^2)\dfrac{dy}{dx} + (2x^2 - 1)y = ax^3,$

$y = ax + cx\sqrt{(1-x^2)}.$

✓ 11. $\cos x \dfrac{dy}{dx} + y - 1 + \sin x = 0,$ $y(\sec x + \tan x) = x + c$

12. $\dfrac{dy}{dx} + \dfrac{1-2x}{x^2} y = 1,$ $\dfrac{y}{x^2} = 1 + ce^{\frac{1}{x}}.$

13. $(1 + y^2)dx = (\tan^{-1} y - x)dy,$

$x = \tan^{-1} y - 1 + ce^{-\tan^{-1} y}.$

Solve, by transformation, the following equations:—

14. $\dfrac{dy}{dx} = mx + ny + p,$

$mnx + n^2y + m + pn = ce^{nx}.$

15. $(x - y^2)dx + 2xy\,dy = 0,$ $\qquad \log x + \dfrac{y^2}{x} = c.$

16. $(x + y)^2 \dfrac{dy}{dx} = a^2,$ $\qquad \dfrac{x + y}{a} = \tan \dfrac{y + c}{a}.$

17. $\dfrac{dy}{dx} + y = xy^3,$ $\qquad \dfrac{1}{y^2} = x + \tfrac{1}{2} + ce^{2x}.$

18. $\dfrac{dy}{dx} = x^3y^3 - xy,$ $\qquad \dfrac{1}{y^2} = x^2 + 1 + ce^{x^2}.$

19. $(1 - x^2)\dfrac{dy}{dx} - xy = axy^2,$ $\qquad \dfrac{1}{y} = c\sqrt{(1 - x^2)} - a.$

20. $3y^2 \dfrac{dy}{dx} - ay^3 = x + 1,$ $\qquad y^3 = ce^{ax} - \dfrac{x + 1}{a} - \dfrac{1}{a^2}.$

21. $x\dfrac{dy}{dx} + y = y^2 \log x,$ $\qquad \dfrac{1}{y} = \log x + 1 + cx.$

22. $\dfrac{dy}{dx} + \dfrac{xy}{2(1 - x^2)} = \dfrac{ax}{y},$ $\qquad y^2 = c\sqrt{(1 - x^2)} - 2a(1 - x^2).$

23. $\dfrac{dy}{dx} = \dfrac{1}{xy + x^2y^3},$ $\qquad \dfrac{1}{x} = 2 - y^2 + ce^{-\frac{1}{2}y^2}.$

CHAPTER III.

EQUATIONS OF THE FIRST ORDER, BUT NOT OF THE FIRST DEGREE.

V.

Decomposable Equations.

38. A differential equation of the first order is a relation between x, y, and p. If the equation is not of the first degree with respect to p, the first step in the solution is usually to solve the equation for p. Suppose, in the first place, that the equation is a quadratic in p; then two values of p in terms of x and y are found. These will generally be irrational functions of x and y; in the exceptional case when they are rational functions, the equation will be decomposable into two equations of the first degree. For example, the equation

$$\left(\frac{dy}{dx}\right)^2 - (x+y)\frac{dy}{dx} + xy = 0 \quad \ldots \quad (1)$$

may be written

$$\left(\frac{dy}{dx} - x\right)\left(\frac{dy}{dx} - y\right) = 0,$$

and is satisfied by putting

$$\frac{dy}{dx} - x = 0 \quad \ldots \ldots \ldots (2)$$

or

$$\frac{dy}{dx} - y = 0. \quad \ldots \ldots \ldots (3)$$

The integrals of these equations are

$$2y = x^2 + c \quad \ldots \ldots \ldots \quad (4)$$

and

$$y = Ce^x \quad \ldots \ldots \ldots \quad (5)$$

respectively. Each of these is therefore an integral of equation (1).

Thus, a decomposable equation of the second degree has two distinct solutions.

Equations Properly of the Second Degree.

39. In a proper, that is, an indecomposable, equation of the second degree, the two expressions for p are the values of a two-valued function of x and y expressed by attaching the ambiguous sign to the radical involved. There is, in this case, but one integral, the ambiguity disappearing in the process of integrating or of rationalizing the result; so that it is, in fact, unnecessary to retain the ambiguous sign in the expression for p. For example, the equation

$$\left(\frac{dy}{dx}\right)^2 = \frac{y}{x} \quad \ldots \ldots \ldots \quad (1)$$

gives $p = \pm\dfrac{\sqrt{y}}{\sqrt{x}}$; whence

$$\frac{dx}{\sqrt{x}} \pm \frac{dy}{\sqrt{y}} = 0;$$

and, integrating,

$$\sqrt{x} \pm \sqrt{y} = \pm\sqrt{c}. \quad \ldots \ldots \ldots \quad (2)$$

But, rationalizing this, we have

$$(x - y)^2 - 2c(x + y) + c^2 = 0, \quad \ldots \quad (3)$$

a single equation for the complete integral.

The system of curves represented by equation (3) consists of parabolas, each of which touches the two axes at the same distance c from the origin, and the different combinations of signs in equation (2) simply correspond to the three arcs into which the parabola is separated by the points of contact.

Systems of Curves corresponding to Equations of Different Degrees.

40. A differential equation of the first degree is, properly speaking, one in which p has a single value for given simultaneous values of x and y. An equation of the second degree is one in which p has two values for given values of x and y, and so on. Thus, such an equation as $p = \sin^{-1} x$ is not an equation of the first degree, because, for any value of x, $\sin^{-1} x$ has an unlimited number of values. The general form of an equation of the first degree is, then,

$$Lp + M = 0,$$

in which L and M denote one-valued functions of x and y.

Two curves of the system cannot, in general, intersect, for in that case there would be two values of p at the point of intersection. The points, if there be any, at which $L = 0$ and $M = 0$, form an exception; for, at these points, p is indeterminate, as exemplified in Art. 19. Thus, the curves of the system either do not intersect at all, or intersect only at certain points where p is indeterminate.* It follows that, in the integral equation, given simultaneous values of x and y must, except in

---

* The same reasoning shows that, the differential equation being of the first degree, points in which two arcs corresponding to the same value of c intersect, in other words, nodes, can only occur at the points where p is indeterminate. Conversely, these points must either be points through which all the curves of the system pass, or else nodes. In the latter case, they may be crunodes through which two real arcs of a particular integral pass, or acnodes through which no arc passes.

the case of the points above mentioned, determine a single value of c, or, at least, values of c which determine a single curve.

For example, the integral of the equation

$$p = 1 + y^2 \qquad \qquad (1)$$

is

$$\tan^{-1} y - x = a. \qquad \qquad (2)$$

If we give particular values to x and y, we find an unlimited number of values of a differing by multiples of π; but, writing the equation in the form

$$y = \tan(x + a),$$

we see that these values determine but a single curve. We, in fact, obtain all the curves of the system by allowing a to range in value from 0 to π; and, as a varies over this range, the curve sweeps over the whole plane once.

If we take the tangent of each member of equation (2), and write $\tan a = c$, we have

$$\frac{y - \tan x}{1 + y \tan x} = c,$$

in which any simultaneous values of x and y determine a single value of c; and c must pass over the range of all real values in order to make the curve sweep once over the entire plane.

41. In general, if the constant of integration is such that different values of it always correspond to different curves, there can be but one value of c for each point; hence the form of the integral is

$$Pc + Q = 0$$

where P and Q are one-valued functions of x and y, and this we may regard as the standard form of the integral. It will be noticed that both $P = 0$ and $Q = 0$ are particular integrals;

the former corresponding to $c = \infty$, and the latter to $c = 0$. Thus, in the example given above, $y = \tan x$ and $y = -\cot x$ are particular integrals.

42. In like manner, the form of the differential equation of the first order and second degree is

$$Lp^2 + Mp + N = 0,$$

where L, M, and N are one-valued functions of x and y. In general, two curves, and two only, representing particular integrals, intersect in a given point. When the expression $Lp^2 + Mp + N$ can be separated into rational factors of the first degree, these curves belong to distinct systems having no connection with one another, as in Art. 38; but, in the general case, they are curves of the same system. Thus, the system of curves representing a proper equation of the second degree is a system of intersecting curves, two curves of the system, in general, passing through a given point. Hence, in the integral equation, given simultaneous values of x and y must generally determine two values of c, or, at least, values of c which determine two and only two curves of the system.

43. Take, for example, the equation

$$p^2 = 1 - y^2, \quad \quad \quad \quad \quad \quad (1)$$

or

$$\frac{dy}{\pm\sqrt{(1 - y^2)}} = dx.$$

The integral is

$$\sin^{-1} y - x = a, \quad \quad \quad \quad (2)$$

or

$$y = \sin(x + a), \quad \quad \quad \quad (3)$$

in which it is permissible to drop the ambiguous sign, because $y = -\sin(x + a)$ may be written $y = \sin(x + \pi + a)$, and is therefore included in the integral (3). Here, as in the example of Art. 40, if we give particular values to x and y, a has an unlimited number of values; for, if θ be the primary value of $\sin^{-1} y$, every expression included in either of the forms

$$2n\pi + \theta \quad \text{or} \quad (2n+1)\pi - \theta,$$

where n is an integer, is a value of $\sin^{-1} x$. These values of a, however, determine but two distinct curves, since values of a differing by a multiple of 2π determine, in (3), the same curve, so that each of the above forms determines but one curve. Equation (3), in fact, represents the system formed by moving the curve of sines, $y = \sin x$, in the direction of the axis of x, and we obtain all the curves of the system while a varies from 0 to 2π, each branch or wave of the curve falling, when $a = 2\pi$, upon the original position of an adjacent branch. In this motion, the curve sweeps twice over that portion of the plane which lies between the straight lines $y = 1$ and $y = -1$, for which portion only the value of p is possible in equation (1).

44. If, in the integral of an equation of the second degree, we so take the constant of integration c, that *different values of it always correspond to different curves of the system*, there can be but two values of c corresponding to a given point. The equation will then take the form

$$Pc^2 + Qc + R = 0$$

where P, Q, and R are one-valued functions of x and y; and this may be regarded as the standard form of the integral.

To reduce equation (3) of the preceding article to the standard form, we have, on expanding,

$$y = \sin x \cos a + \cos x \sin a,$$

in which $\sin a$ and $\cos a$ are to be expressed in terms of a single constant. For this purpose, we do not put $\sin a = c$ and $\cos a = \sqrt{(1 - c^2)}$, because this would require us to introduce an irrelevant factor in rationalizing the equation in c; but we express $\sin a$ and $\cos a$ by the rational expressions the sum of whose squares is unity; that is, we put

$$\sin a = \frac{1 - c^2}{1 + c^2}, \qquad \cos a = \frac{2c}{1 + c^2}.$$

We thus obtain

$$c^2(y + \cos x) - 2c \sin x + y - \cos x = 0,$$

which is the complete integral of equation (1), Art. 43, expressed in the standard form.

Singular Solutions.

45. Representing a set of simultaneous values of x, y, and p by a moving point, every moving point which satisfies a given differential equation is, at each instant, moving in some one of the systems of curves representing the complete integral. In this sense, the latter completely corresponds to the differential equation: nevertheless, there are, in some cases, other curves in which, if a point be moving, it will satisfy the differential equation. For, suppose a curve to exist which, at each of its points, touches one of the curves representing particular integrals; then a point moving in this curve is moving at each instant in the same direction, that is, with the same value of p, as if it were moving in a curve representing a particular integral; hence it satisfies the differential equation.

Such a curve is an *envelope* of the system of curves representing the complete integral, and its equation is called a *singular solution* of the differential equation. An example has already been noticed in Art. 8; the equation

$$xp^2 - yp + a = 0$$

has the singular solution

$$y^2 = 4ax,$$

in addition to the complete integral

$$y = cx + \frac{a}{c}.$$

Now, the latter is the general equation of the tangents to the parabola $y^2 = 4ax$, which accordingly form a system of straight lines of which the singular solution represents the envelope.

46. We shall now show how a singular solution, when it exists, may be found, either directly from the differential equation, or from the complete integral if the latter be known. Two curves of the system touching the envelope at neighboring points intersect in a point which ultimately falls upon the envelope when the two curves are brought into coincidence; hence the envelope is sometimes called the *locus of the intersection of consecutive curves*. While the curves are distinct, two values of p in the differential equation correspond to the point of intersection. These become equal when the curves coincide, that is, when the point is moved up to the envelope; and they become imaginary when the point crosses the envelope. In like manner, if we substitute the coordinates of the point in the integral equation, it determines two values of c while the curves are distinct; and these become equal when the point is moved up to the envelope, and imaginary when the point crosses it. Thus, at every point of the envelope, both the differential equation considered as an equation for p, and the integral equation considered as an equation for c, have a pair of equal roots. Hence, if we form the condition for equal roots in either of these equations, which we shall, for shortness, call the p-equation and the c-equation, we shall have an equation which must be satisfied by every point on the envelope.

47. The expression which vanishes whenever two roots of an equation are equal is called the *discriminant* of the equation. The discriminants of the p-equation and the c-equation are expressions involving x and y. Either of these expressions may break up into factors, the vanishing of any one of which causes the discriminant to vanish. Hence it follows from the preceding article that, if there be a singular solution, its equation is the result of putting the p-discriminant, or one of its factors, equal to zero, and it is likewise the result of putting the c-discriminant, or one of its factors, equal to zero.

For example, in equation (1), Art. 43, the condition for equal roots is evidently $y^2 - 1 = 0$; hence

$$y - 1 = 0 \quad \text{and} \quad y + 1 = 0$$

are the only equations which can possibly be singular solutions. Each of these equations gives, by differentiation, $p = 0$, and is found to satisfy the differential equation. Hence they are singular solutions, the lines they represent being envelopes of the sinusoids represented by the complete integral.

48. The general method of finding the discriminant of an equation is to differentiate it with respect to the unknown quantity and then to eliminate that quantity between the result and the original equation. But, in the case of the equation of the second degree, it is found more simply by solving the equation. Thus the p-equation, in this case, is

$$Lp^2 + Mp + N = 0; \quad \ldots \ldots \quad (1)$$

whence

$$p = \frac{-M \pm \sqrt{(M^2 - 4LN)}}{2L}; \ldots \ldots \quad (2)$$

so that the condition for equal roots is

$$M^2 - 4LN = 0. \ldots \ldots \ldots \quad (3)$$

In like manner, the general form of the c-equation is

$$Pc^2 + Qc + R = 0;$$

and the condition for equal roots is

$$Q^2 - 4PR = 0.$$

For example, in the final equation of Art. 44, the condition for equal roots is
$$4\sin^2 x - 4(y^2 - \cos^2 x) = 0,$$
or
$$1 - y^2 = 0,$$

which is identical with the like condition for the p-equation given in Art. 47.

Cusp-Loci.

49. There are other loci, for points upon which the discriminants vanish, which it is necessary to distinguish from the envelope whose equation alone is a singular solution. There is, in fact, no reason why the values of p derived from the differential equation, when they become equal as the point (x, y) crosses a certain locus, should also become equal to the value of p for a point moving along that locus. Suppose, then, that the two arcs of particular integral curves passing through (x, y) meet, without touching, the locus for which the values of p become equal; and suppose, as will usually be the case, that the values of p become imaginary as we cross the locus; then, when (x, y) is moved up to the locus, the two arcs will come to have a common tangent; and, since they cannot cross the locus, they will form a cusp, becoming branches of the same particular integral curve. Thus, the two values of c which corresponded to the two intersecting arcs will also become identical, and the locus, which is called a *cusp-locus*, is one for which the c-discriminant also vanishes.

For example, the roots of the equation

$$4p^2 = 9x$$

are equal, each being equal to zero, when

$$x = 0;$$

but, since $p = \infty$ for a point moving along this line, this equation does not satisfy the differential equation. The complete integral is

$$(y + c)^2 = x^3,$$

in which the condition of equal roots is $x^3 = 0$. The system of curves is that resulting from moving the semi-cubical parabola $y^2 = x^3$, which has a cusp at the origin, in the direction of the axis of y. This axis is, therefore, a cusp-locus.

Tac-Loci and Node-Loci.

50. In the preceding article, the values of p were supposed to become imaginary as we cross the locus for which they become equal. From equation (2) of Art. 48, it appears that this will be the case if the discriminant changes sign,* but otherwise not; hence, if the factor which vanishes at the locus appears in the p-discriminant with an even exponent, p will not become imaginary in crossing the locus. In this case, the two intersecting arcs cross the locus; and, when (x, y) is moved up to the locus, we shall simply have two particular integrals which touch one another. Such a locus is called a *tac-locus*. Since

---

* Since the discriminant is the product of the squares of the differences of the roots, this will be true also for equations of the third and higher degrees.

the values of c for the two curves remain distinct, the factor indicating a tac-locus does not appear at all in the c-discriminant, but appears in the p-discriminant with an even exponent.*

51. On the other hand, a factor may appear in the c-discriminant with an even exponent, and not at all in the p-discriminant. Through every point of the locus on which such a factor vanishes, the proper number of arcs of particular integral curves pass, but two of them correspond to the same value of c; thus, the point is a node of the curve determined by this value of c, and the locus is called a *node-locus*.

52. The equation
$$xp^2 - (x-a)^2 = 0 \qquad (1)$$

furnishes an example of each of the cases mentioned in the two preceding articles. The complete integral is

$$y + c = \tfrac{2}{3}x^{\tfrac{3}{2}} - 2ax^{\tfrac{1}{2}},$$

or

$$\tfrac{9}{4}(y+c)^2 = x(x-3a)^2. \qquad (2)$$

The p-discriminant is
$$x(x-a)^2, \qquad (3)$$
and the c-discriminant is
$$x(x-3a)^2. \qquad (4)$$

The system of curves is the result of moving the curve

$$\tfrac{9}{4}y^2 = x(x-3a)^2$$

---

* If a squared factor in the p-discriminant satisfies the differential equation, the two arcs of particular integral curves passing through (x, y), instead of crossing the locus when (x, y) is moved up to it, will coincide with it in direction, as in the case of the envelope, Art. 46. But, since p is real on both sides of the locus, the arcs reappear upon the other side of the locus when (x, y) is moved across it. This implies that they coincide with the locus when (x, y) is upon it. Hence, in this case, the squared factor appears also in the c-discriminant, and represents a particular integral.

in the direction of the axis of y. This curve touches the axis of y at the origin, has a node at the point $(3a, 0)$, and, between these points, consists of a loop in which the tangents at the two points where $x = a$ are parallel to one another. Accordingly the factor x, which is common to both discriminants (3) and (4), indicates the envelope $x = 0$; $x - a = 0$ is a tac-locus, and $x - 3a = 0$ is a node-locus.

53. Two values of c become equal, in other words, the c-discriminant vanishes, whenever the point (x, y) is at the ultimate intersection of consecutive curves of the system represented by the c-equation. Suppose this equation to represent a curve having, for all values of c,* one or more nodes or cusps. Considering the intersections of two neighboring curves of the system, it is evident that there are two intersections in the neighborhood of each node, and that these ultimately coincide with the node. Again, there are three (all of which may be real) which ultimately coincide with each cusp. Now, the c-discriminant gives the complete locus of the ultimate intersections: it therefore includes the node-locus repeated twice, and the cusp-locus repeated three times; that is to say, the discriminant contains the factor indicating a node-locus as a squared factor, and it contains the factor indicating a cusp-locus as a cubed factor, as illustrated in the example of Art. 49, where the factor x^3 occurs in the c-discriminant, while the first power only of x occurs in the p-discriminant.

54. A decomposable differential equation of the second degree has no singular solution: for the discriminant is the

---

* If, for a *particular* value of c, a node occurs at the point (x, y), there are no intersections of consecutive curves in its neighborhood, the point does not cause the c-discriminant to vanish, and there are for it the proper number of values for c, and therefore one too many values of p. Hence, at such a point, the p-equation vanishes identically irrespective of the value of p; that is to say, all its coefficients vanish. (See Cayley, *Messenger of Mathematics, New Series*, vol. ii. p. 10.)

If a point cause both the p-equation and the c-equation to vanish identically, it will be a fixed intersection of the curves of the system.

square of the difference between the roots; hence, if the roots are rational, it is the square of a rational function. The systems representing the two complete integrals, in this case, are non-intersecting systems; and the discriminant vanishes only at the tac-locus, at every point of which a curve of one system touches a curve of the other system. Thus, in equation (1), Art. 38, the discriminant is

$$(x+y)^2 - 4xy = (x-y)^2,$$

the square of a rational function; and the line $x = y$ is a tac-locus at every point of which one of the parabolas represented by equation (4) touches one of the exponential curves represented by equation (5).*

Examples V.

Solve the following equations, finding the singular solutions, when they exist, as well as the complete integrals:—

1. $\left(\dfrac{dy}{dx}\right)^2 - a^2y^2 = 0,$ $y = ce^{ax},$ and $y = Ce^{-ax}.$

2. $p(p - y) = x(x + y),$
$\qquad 2y + x^2 = c,$ and $y + x + 1 = Ce^x.$

3. $(x^2 + 1)p^2 = 1,$ $c^2e^{2y} - 2cxe^y = 1.$

---

* In like manner, the discriminant of a decomposable c-equation gives a node-locus. But it is to be noticed that there is no propriety in combining the two integrals of a decomposable p-equation. Thus, if we combine equations (4) and (5) of Art. 38, assuming C and c to be identical, we associate each curve of one system with a particular curve of the other system. But if, before doing this, we change the form of one of the integrals (by introducing a new constant $f(c)$, as explained in Art. 30), we associate the curves differently, and obtain a new result, equally entitled to be considered the integral of the given decomposable differential equation.

§ V.] EXAMPLES. 51

4. $\left(\dfrac{dy}{dx}\right)^2 - \dfrac{a^2}{x^2} = 0,$ $ce^y = x^{\pm a}.$

5. $y^2p^2 = 4a^2,$ $y^2 = c \pm 4ax.$

6. $p^2 - 5p + 6 = 0,$ $y = 2x + c,$ and $y = 3x + C.$

7. $\left(\dfrac{dy}{dx}\right)^2 - \dfrac{a}{x} = 0,$ $(y - c)^2 = 4ax.$

8. $x^2p^2 + 3xyp + 2y^2 = 0,$ $xy = c,$ and $x^2y = C.$

9. $p^3 + 2xp^2 - y^2p^2 - 2xy^2p = 0,$
$y = c,$ $y + x^2 = c,$ and $xy + 1 + cy = 0.$

10. $p^3 - (x^2 + xy + y^2)p^2 + xy(x^2 + xy + y^2)p - x^3y^3 = 0,$
$cy = e^{\frac{1}{2}x^2},$ $cy = 1 + xy,$ and $3y = x^3 + c.$

11. $p^2 + 2py \cot x = y^2,$ $y(1 \pm \cos x) = c.$

12. $\left(\dfrac{dy}{dx}\right)^2 - ax^3 = 0,$ $25(y - c)^2 = 4ax^5.$

13. $x + xp^2 = 1,$ $y = \sqrt{(x - x^2)} + \sin^{-1}\sqrt{x} + c.$

14. $p^2(x^2 + 1)^3 = 1,$ $(y - c)^2 = \dfrac{x^2}{x^2 + 1}.$

15. $y = (x + 1)p^2,$
$c^2 + 2c(x + 1 + y) + (x + 1 - y)^2 = 0.$

16. $yp^2 + 2xp - y = 0,$ $y^2 = 2cx + c^2.$

17. $3xp^2 - 6yp + x + 2y = 0,$ $c^2 + c(x - 3y) + x^2 = 0.$

18. $yp + nx = \sqrt{(y^2 + nx^2)}\sqrt{(1 + p^2)},$
$y + \sqrt{(y^2 + nx^2)} = cx^{1 \pm \sqrt{(1-n-1)}}.$

19. $x^2p^2 - 2xyp + y^2 = x^2y^2 + x^4$, $\quad 2c\dfrac{y}{x} = c^2e^x - e^{-x}$.

20. $3p^2y^2 - 2xyp + 4y^2 - x^2 = 0$,
$$x^2 + y^2 - 4cx + 3c^2 = 0.$$

21. $y^2(1 + p^2) = n^2(x + yp)^2$,
$$(x + c)^2 = (n^2 - 1)y^2 + n^2x^2.$$

VI.

Solution by Differentiation.

55. The differentiation of a differential equation of the first order gives rise to an equation of the second order; but, in the cases now to be considered, the result may be regarded as an equation of the first order, and its integral used in determining that of the given equation.

Let the given equation be solved for y, that is to say, put in the form
$$y = f(x, p); \quad \ldots \ldots \ldots \quad (1)$$

then the result of differentiation will be of the form
$$p = \phi\!\left(x, p, \frac{dp}{dx}\right), \quad \ldots \ldots \ldots \quad (2)$$

which is of the second order as regards y, but, not containing y explicitly, is an equation of the first order between x and p. If, now, we can integrate this equation, we shall have a relation between x, p, and an arbitrary constant. The result of elimi-

§ VI.]   SOLUTION BY DIFFERENTIATION.   53

nating p between this equation and equation (1) will therefore be a relation between x, y, and an arbitrary constant; hence it will be the complete integral required.

56. For example, given the equation

$$\frac{dy}{dx} + 2xy = x^2 + y^2;$$

solving for y, we have

$$y = x + \sqrt{p}; \quad \ldots \ldots \ldots (1)$$

and, differentiating,

$$p = 1 + \frac{1}{2\sqrt{p}}\frac{dp}{dx}. \quad \ldots \ldots \ldots (2)$$

Separating the variables x and p, we have

$$dx = \frac{dp}{2\sqrt{p}(p-1)};$$

and, integrating,

$$x = \tfrac{1}{2} \log \frac{\sqrt{p}-1}{\sqrt{p}+1} + c,$$

or

$$\sqrt{p} = \frac{1 + e^{2x-2c}}{1 - e^{2x-2c}}. \quad \ldots \ldots \ldots (3)$$

Finally, eliminating p between equations (1) and (3), we have the complete integral

$$y = x + \frac{C + e^{2x}}{C - e^{2x}}.$$

57. In attempting this mode of solution, it will sometimes be more advantageous to treat y as the independent variable, and putting p' for $\frac{dx}{dy}$, to derive a differential equation involving y and p'. In either case, the success of the method depends upon our ability to integrate the derived equation. The principal cases in which this can be effected are those in which one of the variables is absent and those in which both variables occur only in the first degree.

It should be noticed that the final elimination of p is frequently inconvenient, or even impracticable; but, when this is the case, we may express x and y in terms of p which then serves as an auxiliary variable.

Equations from which One of the Variables is Absent.

58. If an equation of the first order in which x does not occur explicitly can be solved for p, it takes the directly integrable form

$$\frac{dx}{dy} = f(y), \qquad (1)$$

y being treated as the independent variable. Otherwise let it be solved for y; thus,

$$y = \phi(p); \qquad (2)$$

differentiating,

$$p = \phi'(p)\frac{dp}{dx}, \qquad (3)$$

in which the variables x and p can be separated.

In like manner, an equation not containing y, if not directly integrable, should be put in the form

$$x = \phi(p).$$

Differentiating with respect to y, we have

$$\frac{1}{p} = \phi'(p)\frac{dp}{dy},$$

in which the variables y and p can be separated.

59. As an example, let us take the equation

$$y = p^2 + \tfrac{2}{3}p^3. \quad \ldots \ldots \quad (1)$$

We have, by differentiation,

$$p = (2p + 2p^2)\frac{dp}{dx}, \quad \ldots \ldots \quad (2)$$

which implies either that

$$p = 0, \quad \ldots \ldots \ldots \quad (3)$$

or else that

$$dx = (2 + 2p)dp. \quad \ldots \ldots \quad (4)$$

Eliminating p from equation (1) by means of the first of these, which is not a differential equation for p, we obtain the solution

$$y = 0, \quad \ldots \ldots \ldots \quad (5)$$

which does not contain an arbitrary constant. But, integrating equation (4), we have

$$x + c = 2p + p^2,$$

or

$$p = -1 + \sqrt{(x + c)};$$

and, employing this result to eliminate p from equation (1), we obtain

$$y = \tfrac{1}{3} - x - c + \tfrac{2}{3}(x + c)^{\frac{3}{2}},$$

or, rationalizing,

$$(x + y + c - \tfrac{1}{3})^2 = \tfrac{4}{9}(x + c)^3. \quad \ldots \ldots \quad (6)$$

This equation contains an arbitrary constant, and is the complete integral.

Equation (5), not being a particular case of equation (6), is a singular solution.

60. With respect to an equation of the form

$$y = \phi(p), \quad \ldots \ldots \ldots \quad (1)$$

it may be noticed that

$$y = \phi(0), \quad \ldots \ldots \ldots \quad (2)$$

(which, since ϕ is not necessarily one-valued, may include several equations) is always a solution, for it gives, by differentiation, $p = 0$, and thus satisfies equation (1). The reason of this is readily seen, for the complete integral is capable of expression in the form

$$x = \psi(y) + c, \quad \ldots \ldots \ldots \quad (3)$$

which is the form it would take if derived by direct integration from the form (1), Art. 58; it therefore represents the system of curves which results from moving the curve

$$x = \psi(y)$$

in the direction of the axis of x. If this curve contains points at which $p = 0$, it is evident that the locus of these points, or $y = \phi(0)$, is an envelope; that is, $y = \phi(0)$ is a singular solution.* But, if the point for which $p = 0$ is at an infinite distance, $y = \phi(0)$ will be the particular integral corresponding to $c = \infty$ when the integral is written in the form (3). For

---

* If the p-discriminant were formed, in this case, by the general method (see Art. 48), we should apparently have $\phi'(p) = 0$ as the condition satisfied alike by a singular solution, a cusp-locus, and a tac-locus. But it is to be noticed, that, when $\varphi(p)$ is not a one-valued function, the method may fail to detect a case of equal roots. In fact it is evident, from equation (3), Art. 58, that, if $\phi'(p) = 0$, we must have $\frac{dp}{dx}$ or $\frac{d^2y}{dx^2}$ infinite, which indicates a cusp, except when $p = 0$, which, as we have seen above, gives a singular solution. Thus, a tac-locus does not satisfy $\phi'(p) = 0$. In the example of Art. 59, the roots of $\phi'(p) = 0$ are 0 and -1, $y = \phi(0)$ being the envelope, while $y = \phi(-1) = \frac{1}{3}$ is a cusp-locus.

example, the equation $y = p$ is satisfied by $y = 0$. This is, of course, not a singular solution; but the complete integral is $\log y = x + c$ or $y = Ce^x$, and $y = 0$ is the particular integral corresponding to $c = -\infty$ in the first form, or to $C = 0$ in the second.

Homogeneous Equations.

61. When a homogeneous equation which is not of the first degree can be solved for p, it takes the form

$$\frac{dy}{dx} = f\left(\frac{y}{x}\right)$$

considered in Art. 20. Otherwise it should be put in the form

$$\frac{y}{x} = \phi(p),$$

or

$$y = x\phi(p). \quad \ldots \ldots \ldots (1)$$

Differentiating,

$$p = \phi(p) + x\phi'(p)\frac{dp}{dx}, \quad \ldots \ldots (2)$$

in which the variables can be separated.

62. If p_1 is a root of the equation $p = \phi(p)$,

$$y = p_1 x$$

is always a solution of equation (1); for it gives, by differentiation, $p = p_1$, and substituting these values in equation (1), we have

$$p_1 x = x\phi(p_1),$$

which is satisfied by the hypothesis.

It was shown in Art. 22 that the complete integral, in this case, represents a system of similar curves with the origin as the centre of similitude. It is hence evident that the tangents from the origin to any curve of the system will, if the points of contact be at a finite distance, constitute the envelope of the system; but, if the points of contact be at an infinite distance, they will be asymptotes to the system. In either case, they will be the loci of the points for which $p = \dfrac{y}{x}$ in the differential equation (1), that is to say, for which $p = \phi(p)$; but, in the first case, their equations will be singular solutions;* and, in the second case, they will constitute the particular integral corresponding to $c = 0$ when the complete integral is written in the homogeneous form, as in Art. 22.

Equation of the First Degree in x and y.

63. The equation of the first degree in x and y may be written in the form

$$y = x\phi(p) + f(p). \quad \ldots \ldots (1)$$

Differentiating, we have

$$p = \phi(p) + x\phi'(p)\frac{dp}{dx} + f'(p)\frac{dp}{dx}, \quad \ldots (2)$$

or

$$\frac{dx}{dp} = \frac{\phi'(p)}{p - \phi(p)}x + \frac{f'(p)}{p - \phi(p)}, \quad \ldots (3)$$

which is a linear equation for x regarded as a function of p. The integral gives x as a function of p; the elimination of p is often impracticable, but, in that case, substituting the value of x in equation (1), we have x and y expressed in terms of p as an auxiliary variable.

---

* In this case also, $\phi'(p) = 0$ determines cusp-loci, but fails to detect a tac-locus. See the preceding foot-note.

Clairaut's Equation.

64. The equation

$$y = px + f(p), \quad \ldots \ldots \ldots (1)$$

which is a special case of equation (1) of the preceding article, is known as Clairaut's equation. The result of differentiation is

$$p = p + x\frac{dp}{dx} + f'(p)\frac{dp}{dx},$$

or

$$[x + f'(p)]\frac{dp}{dx} = 0.$$

This equation is satisfied either by putting

$$x + f'(p) = 0, \quad \ldots \ldots \ldots (3)$$

or by putting

$$\frac{dp}{dx} = 0. \quad \ldots \ldots \ldots (4)$$

Equation (3) gives, by the elimination of p from (1), a singular solution; and equation (4) gives, by integration,

$$p = c,$$

whence, from (1),

$$y = cx + f(c). \quad \ldots \ldots \ldots (5)$$

This is the complete integral, as is verified at sight, since $p = c$ is the result of its differentiation.

65. The complete integral, in this case, represents a system of straight lines, and the singular solution a curve to which these lines are tangent. An example has already been noticed in Art. 45. Conversely, every system of straight lines represented by a general equation containing one arbitrary parameter gives rise to a differential equation in Clairaut's form, having, for its singular solution, the equation of the curve to which the system is tangent. We have only to write the equation in the form (5), and to substitute p for the symbol denoting the parameter. For example, the equation of the tangents to the circle

$$x^2 + y^2 = a^2$$

is

$$y = mx + a\sqrt{(1 + m^2)};$$

hence the differential equation is

$$y = px + a\sqrt{(1 + p^2)};$$

or, rationalizing,

$$(x^2 - a^2)p^2 - 2xyp + y^2 - a^2 = 0.$$

Accordingly the condition of equal roots is found to be $x^2y^2 - (x^2 - a^2)(y^2 - a^2) = 0$, or $x^2 + y^2 = a^2$.

66. If we form the condition for equal roots in equation (1), Art. 64, by the general method mentioned in Art. 48, we have to eliminate p from equation (1) by means of its derivative with respect to p; namely,

$$0 = x + f'(p),$$

which is identical with equation (3). In fact, it is obvious that the condition should be the same; for, since the complete integral represents straight lines, there can be neither cusp-locus nor tac-locus. Precisely the same condition expresses the equality of roots in the c-equation, a node-locus being also impossible.

§ VI.] REDUCTION TO CLAIRAUT'S FORM.

67. A differential equation may be reducible to Clairaut's form by a more or less obvious transformation. For example, given the equation

$$y - 2x\frac{dy}{dx} + ay\left(\frac{dy}{dx}\right)^2 = 0;$$

since $d(y^2) = 2y\,dy$, if we multiply through by y, y^2 may be made the dependent variable; thus,

$$y^2 - 2x\frac{y\,dy}{dx} + a\left(\frac{y\,dy}{dx}\right)^2 = 0,$$

or, putting $y^2 = v$,

$$v = x\frac{dv}{dx} - \frac{a}{4}\left(\frac{dv}{dx}\right)^2;$$

hence the integral is

$$y^2 = cx - \tfrac{1}{4}ac^2.$$

EXAMPLES VI.

Solve the following differential equations:—

1. $y = -xp + x^4p^2$, $\qquad\qquad y = \dfrac{c}{x} + c^2$,

   singular solution, $1 + 4x^2y = 0$.

2. $xp^2 - 2yp + ax = 0$, $\qquad\qquad 2y = cx^2 + \dfrac{a}{c}$,

   singular solution, $y^2 = ax^2$.

3. $x + py(2p^2 + 3) = 0$,

   $y = \dfrac{c}{(1+p^2)^{\frac{1}{2}}},\quad x = -\dfrac{cp(2p^2+3)}{(1+p^2)^{\frac{3}{2}}}.$

4. $y = \dfrac{2ap^3}{(p^2+1)^2},\qquad\qquad x = \dfrac{a(p^2-1)}{(p^2+1)^2} + c.$

5. $x + yp = ap^2$,
$$x = \frac{p}{\sqrt{(1 + p^2)}}\{c + a \log [p + \sqrt{(1 + p^2)}]\}.$$

6. $y = (1 + p)x + p^2$, $\quad\begin{cases} x = 2(1 - p) + ce^{-p}, \\ y = 2 - p^2 + ce^{-p}(1 + p). \end{cases}$

7. $y = ap + \sqrt{(1 + p^2)}$,
$$x = a \log [ay + \sqrt{(a^2 + y^2 - 1)}] + \log [y - \sqrt{(a^2 + y^2 - 1)}] + c.$$

8. $2y = xp + \dfrac{a}{p}$,
$$a^2c^2 - 12acxy + 8cy^3 - 12x^2y^2 + 16ax^3 = 0.$$

9. $y = ap + bp^2$,
$$x = a \log [\sqrt{(a^2 + 4by)} - a] + \sqrt{(a^2 + 4by)} + c.$$

10. $a^2yp^2 - 4xp + y = 0$,
$$c^2 + 2cx(3a^2y^2 - 8x^2) - 3a^4x^2y^4 + a^6y^6 = 0.$$

11. $y = xp + \sqrt{(b^2 + a^2p^2)}$, $\qquad y = cx + \sqrt{(b^2 + a^2c^2)}$,
$$\text{singular solution, } \frac{x^2}{a^2} + \frac{y^2}{b^2} = 1.$$

12. $(1 + x^2)p^2 - 2xyp + y^2 - 1 = 0$, $\quad y = cx + \sqrt{(1 - c^2)}$.

13. $y = p(x - b) + \dfrac{a}{p}$, $\qquad$ singular solution, $y^2 = 4a(x - b)$.

14. $ayp^2 + (2x - b)p - y = 0$, $\quad ac^2 + c(2x - b) - y^2 = 0$.

15. $\left(1 - \dfrac{dy}{dx}\right)^2 = e^{-2y} + e^{-2x}\left(\dfrac{dy}{dx}\right)^2$, $\quad e^y = ce^x + \sqrt{(1 + c^2)}$.

16. $x^2(y - px) = yp^2$, $\qquad y^2 = cx^2 + c^2$.

17. $e^{3x}(p - 1) + p^3e^{2y} = 0$, $\qquad e^y = ce^x + c^3$.

18. $(ap^2 - b)xy + (bx^2 - ay^2 + c)p = 0$,

$$y^2 = Cx^2 + \frac{Cc}{b + aC}.$$

19. $\dfrac{2y}{x} - p = f\left(\dfrac{p}{x} - \dfrac{y}{x^2}\right)$,

$y = cx^2 + xf(c)$.

20. $p^3 - 4xyp + 8y^2 = 0$,

$y = c(x - c)^2$.

21. $x^2p^2 - 2(xy - 2)p + y^2 = 0$,

$(y - cx)^2 + 4c = 0$.

22. $y = 2px + y^2p^3$,

$y^2 = cx + \tfrac{1}{8}c^3$.

23. $(px - y)(py + x) = h^2p$,

$y^2 - cx^2 = -\dfrac{ch^2}{c + 1}$.

VII.

Geometrical Applications.

68. The properties of a curve are frequently expressed by means of such magnitudes as the subtangent, the subnormal, the perpendicular from the origin upon the tangent, etc., the general expressions for which involve the coordinates of a point upon the curve together with the value of the derivative at that point. Hence the analytical expression of such a property, or, indeed, of any property which depends upon the tangents to the curve, gives rise to a differential equation. Again, a property relating to an area or volume connected with a curve, or to the length of an arc of the curve, is expressed by a differential equation. Hence the problem to determine the curve having a given property resolves itself into the solution of a differential equation. For example, the expression for the subnormal is yp; hence, to determine the curve whose subnormal is constant and equal to a, we have only to solve the differential equation

$$y\frac{dy}{dx} = a.$$

The integral of this equation is

$$y^2 = 2a(x+c),$$

therefore the curve having the given property is the parabola whose parameter is $2a$, and whose axis is the axis of x, the position of the vertex being indeterminate.

69. The given property is, in some cases, expressed in polar coordinates. Thus, let it be required to determine the curve in which the angle between the radius-vector and the tangent is n times the vectorial angle. Using the expression for the trigonometric tangent of the angle first mentioned, the property is expressed by the equation

$$\frac{rd\vartheta}{dr} = \tan n\theta,$$

or

$$\frac{dr}{r} = \frac{\cos n\theta d\theta}{\sin n\theta}.$$

Integrating,

$$\log r = \frac{1}{n}\log \sin n\theta + C,$$

which may be written in the form

$$r^n = c^n \sin n\theta.$$

The mode in which the constant enters shows, as might have been anticipated, that the several curves which have the property are simply similar curves similarly situated with respect to the pole; thus, when $n = 1$, they are the circles which touch the initial line at the pole.

§ VII.]   *POLAR COORDINATES.*   65

70. As a further illustration, let us consider the curve traced by a point carried by a curve which rolls upon a fixed straight line. By the principle of the instantaneous centre, the straight line joining the carried point with the point of contact of the curve with the fixed line is always normal to the path of the carried point. Considering the carried point as a pole, this line is a radius-vector of the given curve, and the perpendicular from the carried point to the fixed line is the perpendicular from the pole upon a tangent. Denoting these lines by r_1 and p_1, respectively, the nature of the given curve determines a relation between r_1 and p_1. But, taking the fixed line as the axis of x, p_1 is an ordinate of the required curve, and r_1 is the part of the normal intercepted between the point of contact and the axis of x, the expression for which is $y\sqrt{(1 + p^2)}$. The relation between p_1 and r_1 then at once gives the differential equation.

For example, let the parabola $y^2 = 4ax$ roll upon a straight line, and let it be required to determine the curve traced by the focus. The relation between p_1 and r_1, in this case, is

$$p_1^2 = ar_1,$$

therefore the differential equation is

$$y^2 = ay\sqrt{(1 + p^2)},$$

or, solving for p,

$$\frac{dy}{dx} = \frac{\sqrt{(y^2 - a^2)}}{a}.$$

Let us take as the origin the point of the fixed line on which the vertex of the parabola falls in the rolling motion. This determines the constant of integration by the condition that $x = 0$ when $p = 0$, that is to say, when $y = a$. Integrating, we have

$$\int_a \frac{dy}{\sqrt{(y^2 - a^2)}} = \int_0 \frac{dx}{a},$$

or

$$\log \frac{y + \sqrt{(y^2 - a^2)}}{a} = \frac{x}{a},$$

which may be reduced to the form

$$y = \frac{a}{2}\left[e^{\frac{x}{a}} + e^{-\frac{x}{a}}\right] = a \cosh\frac{x}{a}.$$

The curve is the catenary.

71. In another class of examples, the curve required is the singular solution of a differential equation. It is, in this case, frequently possible to write the complete integral at once, and to derive the singular solution from it instead of forming the differential equation. For example, required the curve such that the sum of the intercepts of its tangents upon the axes is constant and equal to a. The equation of the curve is the singular solution of the equation whose complete integral represents the system of lines having the property mentioned. The general equation of this system is

$$\frac{x}{c} + \frac{y}{a - c} = 1,$$

in which c is the arbitrary parameter. Writing it in the form

$$c^2 + c(y - x - a) + ax = 0,$$

the condition of equal roots is

$$(y - x - a)^2 - 4ax = 0,$$

or

$$(y - x)^2 - 2a(x + y) + a^2 = 0,$$

which is the equation of the required curve, and represents a parabola touching the axes at the points $(a, 0)$ and $(0, a)$.

Trajectories.

72. A curve which cuts a system of curves at a constant angle is called a *trajectory* of the system. The case usually considered is that of the *orthogonal trajectory*, which cuts the system of curves at right angles. The differential equation of the trajectory is readily derived from that of the given system of curves; for, at every point of the trajectory, the value of p has a fixed relation to the value of p corresponding to the same values of x and y in the equation of the given system of curves. Denoting the new value of p by p', this relation is, in the case of the orthogonal trajectories,

$$p = -\frac{1}{p'}.$$

If, then, we put $-\dfrac{dx}{dy}$ in place of $\dfrac{dy}{dx}$ in the differential equation of the given system, the result will be the differential equation of the trajectory. The complete integral of this equation will represent a system of curves, each of which is an orthogonal trajectory of the given system. Reciprocally, the curves of the given system are the orthogonal trajectories of the new system.

73. For example, let it be required to determine the orthogonal trajectories of the circles which pass through two given points.

· Taking the straight line which passes through the two given points as the axis of y and the middle point as the origin, and denoting the distance between the points by $2b$, the equation of the given system of circles is

$$x^2 + y^2 + cx - b^2 = 0, \quad \ldots \ldots \quad (1)$$

in which c is the arbitrary parameter. The differential equation derived from this primitive is

$$(x^2 - y^2 + b^2)dx + 2xy\,dy = 0. \quad \ldots \ldots \quad (2)$$

Substituting $-\dfrac{dx}{dy}$ for $\dfrac{dy}{dx}$, we have

$$(y^2 - x^2 - b^2)dy + 2xy\,dx = 0 \quad \ldots \quad (3)$$

for the differential equation of the trajectories. This equation is the same as the result of interchanging x and y in equation (2), except that the sign of b^2 is changed; its integral is therefore

$$x^2 + y^2 + Cy + b^2 = 0; \quad \ldots \quad (4)$$

and the trajectories form a system of circles having the axis of x as the common radical axis, but intersecting it and each other in imaginary points.

74. It is evident that the differential equations of the given system and of the orthogonal trajectories will always be of the same degree, and that, wherever two values of p become equal in the former, the corresponding values of p will be equal in the latter. Hence the loci of equal roots will be the same in each case. Now, the trajectories will meet an envelope of the given system at right angles; and, since the values of p become imaginary in both equations as we cross the envelope, the envelope is a cusp-locus of the system of trajectories. Conversely, a cusp-locus which is, at each point, perpendicular to a curve of the given system, becomes an envelope of the system of trajectories; but every other cusp-locus is also a cusp-locus of the trajectories.

In like manner, a tac-locus of the given system becomes a tac-locus of the trajectories.* A node-locus gives rise to no peculiarity in the system of trajectories.

---

* The case in which the tangent curves of the system cross the tac-locus at right angles forms an exception. In this case, the locus is itself one of the trajectories; and being represented, in the common p-discriminant of the two systems, by a squared factor, we have the case considered in the foot-note on

Examples VII.

1. Determine the curve whose subtangent is n times the abscissa of the point of contact. $\qquad y^n = cx.$

2. Determine the curve whose subtangent is constant, and equal to a. $\qquad ce^x = y^a.$

3. Determine the curve in which the angle between the radius-vector and the tangent is one-half the vectorial angle. $r = c(1 - \cos\theta).$

4. Determine the curve in which the subnormal is proportional to the nth power of the abscissa. $\qquad y^2 = kx^{n+1} + c.$

5. Determine the curve in which the perpendicular upon the tangent from the foot of the ordinate of the point of contact is constant and equal to a, determining the constant of integration in such a manner that the curve shall cut the axis of y at right angles.

$$\text{The catenary } y = a \cosh\frac{x}{a}.$$

---

page 48. For example, the tac-locus $x = a$ in Art. 52 is perpendicular to the system of curves representing the complete integral; the equation of the trajectories is

$$(x - a)^2 p^2 - x = 0, \quad \ldots \ldots \ldots \ldots \quad (1)$$

of which the integral is

$$y + C = 2\sqrt{x} + \sqrt{a} \log \frac{\sqrt{a} - \sqrt{x}}{\sqrt{a} + \sqrt{x}}. \quad \ldots \ldots \ldots \quad (2)$$

The system is that which results from moving the curve

$$y = 2\sqrt{x} + \sqrt{a} \log \frac{(\sqrt{a} - \sqrt{x})^2}{a - x}$$

in the direction of the axis of y. This curve is symmetrical to the axis of x since $\sqrt{x}$ admits of a change of sign, and it has a cusp at the origin, so that the axis of y is a cusp-locus. The line $x = a$ is an asymptote which is approached by branches on both sides of it; and the result of putting $C = \infty$ in equation (2) is, in fact, this line, or rather the line doubled, for, if C is infinite, we must, in order to have y finite, put $x = a$.

6. Determine the curve in which the perpendicular from the origin upon the tangent is equal to the abscissa of the point of contact.
$$x^2 + y^2 = 2cx.$$

7. Determine the curve such that the area included between the curve, the axis of x, and an ordinate, is proportional to the ordinate.
$$y^n = ce^x.$$

8. Determine the curve in which the portion of the axis of x intercepted between the tangent and the normal is constant, and interpret the condition of equal roots for p.
$$2(x - c) = a\log[a \pm \sqrt{(a^2 - 4y^2)}] \mp \sqrt{(a^2 - 4y^2)}.$$

9. Determine the curve such that the area between the curve, the axis of x and two ordinates is proportional to the corresponding arc.
$$y = \cosh\frac{x}{a}.$$

10. Determine the curve in which the part of the tangent intercepted by the axes is constant.
$$x^{\frac{2}{3}} + y^{\frac{2}{3}} = a^{\frac{2}{3}}.$$

11. Determine the curve in which a and β being the intercepts upon the axes made by the tangent $ma + n\beta$ is constant.
$$\text{The parabola } (ny - mx)^2 - 2a(ny + mx) + a^2 = 0.$$

12. Determine the curve in which the area enclosed between the tangent and the coordinate axes is equal to a^2.
$$\text{The hyperbola } 2xy = a^2.$$

13. Determine the curve in which the projection upon the axis of y of the perpendicular from the origin upon a tangent is constant, and equal to a.
$$\text{The parabola } x^2 = 4a(a - y).$$

14. Determine the curve in which the abscissa is proportional to the square of the arc measured from the origin.
$$\text{The cycloid } y = a\sin^{-1}\frac{\sqrt{x}}{\sqrt{a}} + \sqrt{(ax - x^2)}.$$

15. Determine the orthogonal trajectories of the hyperbolas $xy = a$.
$$\text{The hyperbolas } x^2 - y^2 = c.$$

16. Determine the orthogonal trajectories of the parabolas $y^2 = 4ax$.
The ellipses $2x^2 + y^2 = c^2$.

17. Determine the orthogonal trajectories of the parabolas of the nth degree $a^{n-1}y = x^n$. $\qquad ny^2 + x^2 = c^2$.

18. Find the orthogonal trajectories of the confocal and coaxial parabolas $y^2 = 4a(x + a)$. $\qquad$ The system is self-orthogonal.

19. Show generally that a system of confocal conics is self-orthogonal.

20. Find the orthogonal trajectories of the ellipses $\dfrac{x^2}{a^2} + \dfrac{y^2}{b^2} = 1$ when a is constant and b arbitrary. $\qquad x^2 + y^2 = 2a^2 \log x + c$.

21. Find the orthogonal trajectories of the cardioids $r = a(1 - \cos\theta)$.
$r = c(1 + \cos\theta)$.

22. Determine the orthogonal trajectories of the similar ellipses $\dfrac{x^2}{a^2} + \dfrac{y^2}{b^2} = n^2$, n being the arbitrary parameter. $\qquad y^{b^2} = cx^{a^2}$.

23. Find the orthogonal trajectories of the ellipses $\dfrac{x^2}{a^2} + \dfrac{y^2}{b^2} = 1$ when $\dfrac{1}{a^2} + \dfrac{1}{b^2} = \dfrac{1}{k^2}$. $\qquad (xy)^{2k^2} = ce^{x^2 + y^2}$.

24. Find the orthogonal trajectories of the system of curves $r^n \sin n\theta = a^n$. $\qquad r^n \cos n\theta = c^n$.

25. Find the orthogonal trajectories of the curves $r = \log\tan\theta + a$.
$\dfrac{2}{r} = \sin^2\theta + c$.

CHAPTER IV.

EQUATIONS OF THE SECOND ORDER.

VIII.

Successive Integration.

75. We have seen, in Chapter I., that the complete integral of a differential equation of the second order must contain two arbitrary constants, and that it is the primitive from which the given differential equation might have been derived by differentiating twice and using the results to eliminate the constants. The order in which the differentiations and eliminations take place is evidently immaterial; for, denoting the constants by c_1 and c_2, and the first and second derivatives of y by p and q, all the equations which can arise in the process form a consistent system of relations between x, y, c_1, c_2, p, and q, and these are equivalent to three independent algebraic relations between these six quantities. If, after differentiating the primitive, we eliminate the constant c_2, the result will be a relation between x, y, c_1, and p, that is to say, a differential equation of the first order; and, if we further differentiate this equation, and eliminate c_1, the result will be the differential equation of the second order. Now, regarding the latter as given, the relation between x, y, c_1, and p is called a *first integral*; and the complete integral, or relation between x, y, c_1, and c_2, is also the complete integral of this first integral, c_2 being the constant introduced by the second integration.

76. As an illustration, let the given equation be

$$\frac{d^2y}{dx^2} + y = 0. \quad \ldots \ldots \ldots (1)$$

If this be multiplied by $2p$, it becomes

$$2p\frac{dp}{dx} + 2y\frac{dy}{dx} = 0; \quad \ldots \ldots (2)$$

and, since this equation is the result of differentiating

$$p^2 + y^2 = c^2. \quad \ldots \ldots \ldots (3)$$

(the constant, which is, for convenience, denoted by c^2, disappearing in the differentiation), equation (3) is a first integral of equation (1). It may be written

$$\frac{dy}{\sqrt{(c^2 - y^2)}} = dx;$$

and its integral, which is

$$\sin^{-1}\frac{y}{c} = x + a,$$

or

$$y = c\sin(x + a), \quad \ldots \ldots (4)$$

where a is a second constant of integration, is the complete integral of equation (1). Expanding $\sin(x + a)$, and putting

$$A = c\cos a, \quad B = c\sin a,$$

the complete integral may also be written in the form

$$y = A\sin x + B\cos x, \quad \ldots \ldots (5)$$

in which A and B are the two arbitrary constants.

The First Integrals.

77. It is shown, in Arts. 14 and 15, that a differential equation of the second order represents a doubly infinite system of curves. In fact, if, in the complete integral, we attribute a fixed value to one of the constants, we have a singly infinite system; and, therefore, corresponding to different values of this constant, we have an unlimited number of such systems. For example, if, in the complete integral (4) of the preceding article, we regard c as a fixed constant, the equation represents a system of equal sinusoids each having the axis of x for its axis and c for the value of its maximum ordinate, but having points of intersection with the axis depending upon the arbitrary constant a. The first integral (3) is the differential equation of this system; and equation (1), which does not contain c, represents all such systems obtained by varying the value of c.

On the other hand, if, in equation (4), we regard a as fixed, we have a system of sinusoids cutting the axis in fixed points, but having maximum ordinates depending upon the constant c, which is now regarded as arbitrary. If now we differentiate equation (4) and eliminate c, we have the differential equation of this system, namely,

$$y = p \tan(x + a), \quad \ldots \ldots \ldots (6)$$

which, being a relation between x, y, p and a constant, is another first integral of equation (1). The result of eliminating p between the first integrals (3) and (6) would, of course, be the complete integral (4).

78. Consider now the form (5) of the complete integral. If we regard A as fixed, the singly infinite system represented is one selected in still another manner from the doubly infinite system; it consists, in fact, of those members of the doubly infinite system which pass through the point ($\tfrac{1}{2}\pi$, A). The

§ VIII.] *THE FIRST INTEGRALS.* 75

differential equation of this system, which is found by differentiating, and eliminating B, is

$$y \sin x + p \cos x = A, \quad \ldots \ldots \quad (7)$$

which is, accordingly, another first integral of equation (1) Again, regarding B as fixed, and eliminating A from equation (5), we obtain the first integral

$$y \cos x - p \sin x = B. \quad \ldots \ldots \quad (8)$$

In like manner, to every constant which may be employed as a parameter in expressing the general equation of the doubly infinite system of curves there corresponds a first integral of the differential equation of the second order. Thus, the number of first integrals is unlimited.

79. If c_1 and c_2 are two *independent* parameters, that is to say, such that one cannot be expressed in terms of the other, all the other parameters may be expressed in terms of these two. Accordingly, the two first integrals which correspond to c_1 and c_2, which may be put in the form

$$f_1(x, y, p) = c_1, \quad f_2(x, y, p) = c_2,$$

may be regarded as two *independent* first integrals from which all the first integrals may be derived. For example, if the first integrals (7) and (8) of the preceding article be regarded as the two independent first integrals, equation (3) of Art. 76 may be derived from them by squaring and adding, because $c^2 = A^2 + B^2$.

It must be remembered that no two first integrals are independent when regarded as differential equations of the first order ; for they must both give rise, by differentiation, to the same equation of the second order. They are only independent in the sense that the constants involved are independent, so that they may be regarded as independent *algebraic* relations between the five quantities x, y, p, c_1, and c_2, from which, by the elimination of p, the relation between x, y, c_1, and c_2 can be found independently of the differential relation between x, y, and p.

Integrating Factors.

80. If a first integral of a given differential equation of the second order be put in the form $f(x, y, p) = c$ and differentiated, the result, not containing c, will be a relation between x, y, p, and q, which is satisfied by every set of simultaneous values of these quantities which satisfies the given differential equation. This result will therefore either be the given equation, or else the product of that equation by a factor which does not contain q. In the first case, the given equation is said to be an *exact differential equation;* in the latter, the factor which makes it exact is called an *integrating factor.* In general, to every first integral there corresponds an integrating factor. For example, differentiating equations (7) and (8) of Art. 78, we find the corresponding integrating factors of the equation

$$\frac{d^2y}{dx^2} + y = 0$$

to be $\cos x$ and $\sin x$ respectively. Again, the integrating factor p was employed, in Art. 76, in finding the first integral (3) by means of which we solved the equation.

81. It is to be noticed that an exact equation formed, as in the case last mentioned, by means of an integrating factor containing p, is really a decomposable equation consisting of the given differential equation of the second order and the differential equation of the first order which results from putting the integrating factor equal to zero. The exact differential equation therefore represents, in this case, not only the doubly infinite system, but also a singly infinite system which does not satisfy the given differential equation. This system consists of the singular solutions of the several singly infinite systems represented by the first integral when different values are given to the constant contained in it. For example, equation (2), Art. 76, is satisfied by $y = C$, which does not satisfy equation (1), but is the solution of $p = 0$; accordingly, the first integral (3) has the singular solutions $y = \pm c$, which, when c is arbitrary, form the singly infinite system of straight lines parallel to the axis of x. In fact, a singular solution of a first integral represents a line, which, at each of its points, touches a particular curve of the doubly infinite system. The values of x, y, and p, for a point moving in such a line, are therefore the same as for a point moving in a particular integral curve; but the values of q are, in general, different;* hence such a point does not satisfy the given differential equation.

---

* The values of q will, however, be the same if the line in question has at every point the same curvature as the particular integral curve which it touches at that point; and its equation will then be a singular solution. The case is analogous to that of the singular solution of an equation of the first order; the given equation being supposed of a degree higher than the first in q, and a necessary (but not a sufficient) condition being that two values of q shall become equal for the values of x, y, and p in question. Suppose, for example, the doubly infinite system of curves represented by the differential equation to consist of all the circles whose centres lie upon a fixed curve. In order to determine the particular integrals which pass through an assumed point (x, y) in the direction determined by an assumed value of p, we must draw a straight line through (x, y) perpendicular to the assumed direction, the required particular integrals being circles whose centres are the points where this line cuts the fixed curve. These circles correspond to the several values of q which are consistent with the assumed values of x, y, and p. When the line touches the fixed curve, two of the values of q are equal, and the values of x, y, and p satisfy the condition of equal roots in the differential equation considered as an equation for q. Consider now an involute of the fixed curve; its normals touch the given curve; hence the values of x, y, and p, at any of its points, satisfy the condition of equal roots. Now, the circle corresponding to the twofold value of q is the circle of curvature of the involute, so that the value of q for a point moving in the involute is the same as its value for a point moving in a particular integral curve, and the equation of the involute is a singular solution. Thus the involutes of the fixed curve constitute a singly infinite system of singular solutions, and the relation between x, y, and p, which is satisfied

Derivation of the Complete Integral from Two First Integrals.

82. It sometimes happens that it is easier to obtain two independent first integrals than to effect the integration of one of the first integrals. The elimination of p between the two first integrals then gives the complete integral. For example, as an obvious extension of the results obtained in Art. 80, we see that both $\cos ax$ and $\sin ax$ are integrating factors of the equation

$$\frac{d^2y}{dx^2} + a^2y = 0;$$

and, since these expressions contain x only, they are also integrating factors of the more general equation:—

$$\frac{d^2y}{dx^2} + a^2y = X \quad \ldots \quad \ldots \quad (1)$$

if X is a function of x only. Thus, we have the exact differential equation,

$$\cos ax \frac{d^2y}{dx^2} + a^2 y \cos ax = X \cos ax,$$

and its integral, which is

$$\cos ax \frac{dy}{dx} + ay \sin ax = \int X \cos ax\, dx + c_1 \ldots \quad (2)$$

is a first integral of equation (1). In like manner, the integrating factor $\sin ax$ leads to the first integral

$$\sin ax \frac{dy}{dx} - ay \cos ax = \int X \sin ax\, dx - c_2. \ldots \quad (3)$$

---

by all the involutes (in other words, their differential equation) satisfies the condition of equal roots; that is to say, it is the result of equating to zero the discriminant of the q-equation or one of its factors.

§ VIII.] *ELIMINATION OF p FROM TWO FIRST INTEGRALS.* 79

Eliminating p between equations (2) and (3), we have

$$ay = \sin ax \int X \cos ax dx - \cos ax \int X \sin ax dx + c_1 \sin ax + c_2 \cos ax,$$

the complete integral of equation (1).

83. The principle of this method has already been applied to the solution of equations of the first order in Art. 55. The method there explained, in fact, consists in forming the equation of the second order of which the given equation is a first integral, then finding an independent first integral, and deriving the complete integral by the elimination of p. But it is to be noticed that the given equation, containing, as it does, no arbitrary constant, is only a particular case of the first integral of the equation of the second order corresponding to a particular value of the constant which should be contained in it. Accordingly, the final equation is the result of giving the same particular value to this constant in the complete integral of the equation of the second order. For example, in the solution of Clairaut's equation, Art. 64, the equation of the second order is $\frac{d^2y}{dx^2} = 0$; the first integral, of which the given equation is a special case, is $y + C = xp + f(p)$; and the complete integral is $y + C = cx + f(c)$, which represents all straight lines; whereas the required result is the singly infinite system of straight lines corresponding to $C = 0$.*

---

* In accordance with Art. 81, it would seem that a singular solution of the given equation, when it exists, could not satisfy the equation of the second order, and therefore must correspond to a factor which divides out, just as $x + f'(p)$ does in the solution of Clairaut's equation. This is indeed true when the singular solution belongs to the generalized first integral, as in this case it does to $y + C = cx + f(c)$. But generally the singular solution belongs only to the given equation; and there is no reason why a singular solution of a *particular* first integral should not satisfy the differential equation of the second order. Thus a singular solution does not generally present itself in the process of "solution by differentiation," as it does in the case of Clairaut's equation.

Exact Differential Equations of the Second Order.

84. An exact differential equation of the second order is the result of differentiating a first integral in the form

$$f(x, y, p) = c. \quad \ldots \ldots \quad (1)$$

Hence it will be of the form

$$\frac{df}{dx} + \frac{df}{dy}p + \frac{df}{dp}q = 0, \quad \ldots \ldots \quad (2)$$

in which the partial derivatives $\frac{df}{dx}$, $\frac{df}{dy}$ and $\frac{df}{dp}$, are functions of x, y and p; so that the latter forms the entire coefficient of q in the equation. Hence, if a given equation of the second order is exact, we can, from this coefficient, find, by integration with respect to p, the form of the function f so far as it depends upon p; that is to say, we can find all the terms of the integral which contain p. These terms being found, their complete derivative must be subtracted from the first member of the given differential equation, and the remainder, which will be a differential expression of the first order, must be examined. If this remainder is exact, the whole expression is evidently exact; and its integral is the sum of the terms already found and the integral of the remainder.

85. As an illustration, let the given equation be

$$(1 - x^2)\frac{d^2y}{dx^2} - x\frac{dy}{dx} + y = 0. \quad \ldots \ldots \quad (1)$$

The terms containing q are $(1 - x^2)\frac{dp}{dx}$; and, integrating this with respect to p, we have $(1 - x^2)p$ for the part of the integral

§ VIII.] *EXACT EQUATIONS.* 81

which contains p. The complete derivative of this expression is

$$(1 - x^2)\frac{d^2y}{dx^2} - 2x\frac{dy}{dx};$$

and, subtracting this from the first member of equation (1), we have the remainder

$$x\frac{dy}{dx} + y = 0,$$

which is the derivative of xy. Hence equation (1) is exact, and its integral is

$$(1 - x^2)p + xy = c_1. \quad \ldots \ldots (2)$$

Again, if we multiply equation (1) by p, it becomes

$$(1 - x^2)p\frac{dp}{dx} - xp^2 + yp = 0. \ldots \ldots (3)$$

In this form, the integral of the terms containing q is $\frac{1}{2}(1 - x^2)p^2$, of which the complete derivative is

$$(1 - x^2)p\frac{dp}{dx} - xp^2.$$

The remainder, in this case, is yp, which is the exact derivative of $\frac{1}{2}y^2$; hence equation (3) is also exact, and its integral is

$$(1 - x^2)p^2 + y^2 = c_2. \quad \ldots \ldots (4)$$

Equations (2) and (4) are two first integrals of equation (1); hence, eliminating p, we have the complete integral

$$c_1^2 - 2c_1xy + y^2 - c_2(1 - x^2) = 0, \ldots \ldots (5)$$

which represents a system of conics having their centres at the origin, and touching the straight lines $x = \pm 1$.

Equations in which y does not occur.

86. A differential equation of the nth order which does not contain y is equivalent to an equation of the $(n-1)$th order for p. The value of p as a function of x obtained by integrating this will contain $n-1$ constants; and the remaining constant will appear in the final integration, which will take the form

$$y = \int p\,dx + C.$$

If the given equation is of the first degree with respect to the derivatives, it will be a linear equation because the coefficients do not contain y. Thus, if the equation is of the second order, it may be put in the form

$$\frac{d^2y}{dx^2} + f(x)\frac{dy}{dx} = \phi(x),$$

or

$$\frac{dp}{dx} + Pp = Q,$$

a linear equation of the first order for p. For example, the equation

$$(1 + x^2)\frac{d^2y}{dx^2} + x\frac{dy}{dx} + ax = 0$$

is equivalent to

$$\frac{dp}{dx} + \frac{x}{1 + x^2}p = -\frac{ax}{1 + x^2}.$$

The integral of this is

$$p = -a + \frac{c_1}{\sqrt{(1 + x^2)}};$$

and, integrating again,

$$y = c_2 - ax + c_1 \log[x + \sqrt{(1 + x^2)}].$$

87. In general, an equation of the nth order which does not contain y, and in which the lowest derivative is of the rth order, is equivalent to an equation of the $(n - r)$th order for the determination of this derivative. For example,

$$\frac{d^4y}{dx^4} = a^2 \frac{d^2y}{dx^2}$$

is equivalent to

$$\frac{d^2q}{dx^2} = a^2 q.$$

Integrating, we have

$$q = \frac{d^2y}{dx^2} = c_1 e^{ax} + c_2 e^{-ax};$$

and, integrating twice more,

$$y = Ae^{ax} + Be^{-ax} + Cx + D.$$

Equations in which x does not occur.

88. An equation of the second order in which x does not occur may be reduced to an equation of the first order between y and p by putting

$$\frac{d^2y}{dx^2} = \frac{dp}{dx} = \frac{dp}{dy}\frac{dy}{dx} = p\frac{dp}{dy}.$$

For example, the equation

$$y^2 \frac{d^2y}{dx^2} = \left(\frac{dy}{dx}\right)^3. \quad \ldots \ldots \ldots (1)$$

thus becomes

$$y^2 p \frac{dp}{dy} = p^3, \quad \ldots \ldots \ldots (2)$$

or

$$\frac{dp}{p^2} = \frac{dy}{y^2};$$

whence
$$\frac{1}{p} = \frac{1}{y} + c_1,$$
or
$$dx = \frac{dy}{y} + c_1 dy;$$
and, integrating again,
$$x = \log y + c_1 y + c_2. \quad \ldots \ldots \quad (3)$$

In equation (2), we rejected the solution $p = 0$, which gives $y = C$; but it is to be noticed that the equation is still satisfied by $p = 0$ after the rejection of the factor p; accordingly, $y = C$ is a particular system of integrals included in the complete integral (3), as will be seen by writing the latter in the form
$$y = A + B(x - \log y),$$
and making $B = 0$.

89. If the equation contains higher derivatives, they may, in like manner, be expressed in terms of derivatives of p with respect to y. Thus,
$$\frac{d^3y}{dx^3} = \frac{d}{dx}\frac{d^2y}{dx^2} = p\frac{d}{dy}\left(p\frac{dp}{dy}\right) = p^2\frac{d^2p}{dy^2} + p\left(\frac{dp}{dy}\right)^2.$$

In like manner, the expression for the fourth derivative may be found by applying the operation $p\dfrac{d}{dy}$ to this last result, and so on.

The Method of Variation of Parameters.

90. When the solution of an equation in which the second member is zero is known in the form $y = f(x)$, the more general equation in which the second member is a function of x may sometimes be solved by assuming the value of y in

§ VIII.] *THE METHOD OF VARIATION OF PARAMETERS.* 85

the same form as that which satisfies the simpler equation, except that the constants or parameters in that solution are now assumed to be variables. By substituting for y in the given equation its assumed value, we obtain an equation which must be satisfied by these new variables. When the given equation is of the first order, there is but one new variable, and the method amounts merely to a transformation of the dependent variable; but when the equation is of the nth order, the assumption involves n new variables, and we are at liberty to impose $n - 1$ other conditions upon them beside the condition that the given equation shall be satisfied. The conditions which produce the simplest result are that the derivatives of y, of all orders lower than the nth, shall have the same values when the parameters are variable as when they are constant.

91. For example, given the equation

$$\frac{d^2y}{dx^2} + a^2y = X, \quad \ldots \ldots \quad (1)$$

we assume

$$y = C_1 \cos ax + C_2 \sin ax, \ldots \ldots \quad (2)$$

which, if C_1 and C_2 are constant, satisfies the equation when $X = 0$. Now, if C_1 and C_2 are variable, we may assume this value of y to satisfy equation (1), and, at the same time, impose a second condition upon the two new variables. Differentiating, we have

$$\frac{dy}{dx} = -aC_1 \sin ax + aC_2 \cos ax + \frac{dC_1}{dx} \cos ax + \frac{dC_2}{dx} \sin ax,$$

in which the first two terms form the value of $\frac{dy}{dx}$ when C_1 and C_2 are constant. We now assume, as the second condition mentioned above,

$$\frac{dC_1}{dx} \cos ax + \frac{dC_2}{dx} \sin ax = 0, \ldots \ldots \quad (3)$$

which makes

$$\frac{dy}{dx} = -aC_1 \sin ax + aC_2 \cos ax.$$

Differentiating again, we have

$$\frac{d^2y}{dx^2} = -a^2 C_1 \cos ax - a^2 C_2 \sin ax - a\frac{dC_1}{dx} \sin ax + a\frac{dC_2}{dx} \cos ax.$$

Substituting in equation (1), we obtain

$$- a\frac{dC_1}{dx} \sin ax + a\frac{dC_2}{dx} \cos ax = X \quad \ldots \quad (4)$$

as the condition that y, in equation (2), shall satisfy the given equation. Equations (3) and (4) give, by elimination,

$$- a\frac{dC_1}{dx} = X \sin ax, \quad a\frac{dC_2}{dx} = X \cos ax;$$

whence

$$C_1 = -\frac{1}{a}\int X \sin ax\,dx + c_1, \quad C_2 = \frac{1}{a}\int X \cos ax\,dx + c_2;$$

and, substituting in equation (2),

$$y = -\frac{1}{a}\cos ax \int X \sin ax\,dx + \frac{1}{a}\sin ax \int X \cos ax\,dx$$
$$+ c_1 \cos ax + c_2 \sin ax,$$

as otherwise found in Art. 82.

The method of variation of parameters is of historic interest as one of the earliest general methods employed. It may occasionally be applied also when the term neglected in finding the form to be assumed for the value of y is not a mere function of x; but, for the most part, examples which can be solved by it can be more readily solved by the methods given in the succeeding chapters.

Examples VIII.

Solve the following differential equations :—

1. $\dfrac{d^2y}{dx^2} = xe^x$, $\qquad y = (x-2)e^x + c_1x + c_2$.

2. $\dfrac{d^3y}{dx^3} = \sin^3 x$, $\quad y = \tfrac{7}{9}\cos x - \tfrac{1}{27}\cos^3 x + c_1x^2 + c_2x + c_3$.

3. $x^3\dfrac{d^3y}{dx^3} = 1$, $\qquad y = \tfrac{1}{2}\log x + c_1x^2 + c_2x + c_3$.

4. Find a first integral of $\dfrac{d^2x}{dt^2} = f(x)$, $\quad \left(\dfrac{dx}{dt}\right)^2 = 2\displaystyle\int f(x)\,dx$.

5. $\dfrac{d^2y}{dx^2} = ax + by \;[b>0]$, $\quad -ax + by = Ae^{x\sqrt{b}} + Be^{-x\sqrt{b}}$.

6. $\dfrac{d^2y}{dx^2} = ax - by \;[b>0]$,
$\qquad\qquad ax - by = A\sin x\sqrt{b} + B\cos x\sqrt{b}$.

7. $\dfrac{d^2y}{dx^2} = e^y$, $\qquad \dfrac{\sqrt{(2e^y+c^2)}-c}{\sqrt{(2e^y+c^2)}+c} = Ce^{cx}$,

$\qquad\qquad e^{\frac{1}{2}y} = \dfrac{\sqrt{2}}{C-x}$, or

$\qquad\qquad 2e^y = c^2\sec^2(\tfrac{1}{2}cx + C)$,

according as the first constant of integration is c^2, 0, or $-c^2$.

8. $a^2\left(\dfrac{d^2y}{dx^2}\right)^2 = 1 + \left(\dfrac{dy}{dx}\right)^2$, $\quad y + c_2 = a\cosh\dfrac{x+c_1}{a}$.

9. $\dfrac{d^2y}{dx^2} = \left(\dfrac{dy}{dx}\right)^2 + 1$, $\qquad c_2e^y = \cos(x+c_1)$.

10. $\dfrac{d^2y}{dx^2} + \dfrac{1}{x}\dfrac{dy}{dx} = 0,$ $y = c_1 \log x + c_2.$

11. $\dfrac{d^2y}{dx^2} + y\dfrac{dy}{dx} = 0,$ $\dfrac{a+y}{a-y} = e^{a(x+b)}.$

12. $\dfrac{d^2y}{dx^2} = x\dfrac{dy}{dx},$ $y = c_1 \int e^{\frac{1}{2}x^2} dx + c_2.$

13. $\dfrac{d^2y}{dx^2} + \left(\dfrac{dy}{dx}\right)^2 + 1 = 0,$ $y = \log \sin(x - a) + \beta.$

14. Show that $y\dfrac{d^2x}{dt^2} - x\dfrac{d^2y}{dt^2}$ is an exact differential.

15. $\dfrac{d^2y}{dx^2} = \dfrac{2y}{x^2},$ $y = \dfrac{c_1}{x} + c_2 x^2.$

16. $(1 - x^2)\dfrac{d^2y}{dx^2} - x\dfrac{dy}{dx} = 2,$ $y = (\sin^{-1}x)^2 + c_1 \sin^{-1}x + c_2.$

17. $(1 - x^2)\dfrac{d^2y}{dx^2} + x\dfrac{dy}{dx} = ax,$
$y = ax + c_1[\sin^{-1}x + x\sqrt{(1 - x^2)}] + c_2.$

18. $y\dfrac{d^2y}{dx^2} + \left(\dfrac{dy}{dx}\right)^2 = 1,$ $y^2 = x^2 + c_1 x + c_2.$

19. $x\dfrac{d^2y}{dx^2} + \dfrac{dy}{dx} = x^n,$ $y = \dfrac{x^{n+1}}{(n+1)^2} + c_1 \log x + c_2.$

20. $(1 + x^2)\dfrac{d^2y}{dx^2} + 1 + \left(\dfrac{dy}{dx}\right)^2 = 0,$
$y = c_2 - \dfrac{x}{c_1} + \dfrac{c_1^2 + 1}{c_1^2} \log(1 + c_1 x).$

§ VIII.]   EXAMPLES.   89

21. $\dfrac{d^2y}{dx^2} = \dfrac{1}{a\sqrt{y}}$, $x = \dfrac{2a^{\frac{1}{2}}}{3}(y^{\frac{1}{2}} - 2c_1)(c_1 + y^{\frac{1}{2}})^{\frac{1}{2}} + c_2$.

22. $y(1 - \log y)\dfrac{d^2y}{dx^2} + (1 + \log y)\left(\dfrac{dy}{dx}\right)^2 = 0$,

$$\log y = 1 + \dfrac{1}{c_1 x + c_2}.$$

23. $y\dfrac{d^2y}{dx^2} - \left(\dfrac{dy}{dx}\right)^2 = y^2 \log y$, $\log y = c_1 e^x + c_2 e^{-x}$.

24. $(a^2 - x^2)\dfrac{d^2y}{dx^2} - \dfrac{a^2}{x}\dfrac{dy}{dx} + \dfrac{x^2}{a} = 0$,

$$y = c_1\sqrt{(a^2 - x^2)} + \dfrac{x^2}{2a} + c_2.$$

25. $\dfrac{dy}{dx} = x\dfrac{d^2y}{dx^2} + \sqrt{\left[1 + \left(\dfrac{d^2y}{dx^2}\right)^2\right]}$,

$$y = \tfrac{1}{2}c_1 x^2 + x\sqrt{(1 + c_1^2)} + c_2.$$

26. $\dfrac{d^2u}{d\theta^2} + u = \dfrac{1}{(1 + k^2 \sin^2\theta)^{\frac{3}{2}}}$,

$$u = \dfrac{\sqrt{(1 + k^2 \sin^2\theta)}}{1 + k^2} + e\cos(\theta - a).$$

27. Determine the curve in which the normal is equal to the radius of curvature, but in the opposite direction.

$$\text{The catenary } y = c \cosh\dfrac{x}{c}.$$

28. Determine the curve in which the radius of curvature is double the normal, and in the same direction.

$$\text{The cycloid } -x = c\sin^{-1}\dfrac{c - y}{c} + \sqrt{(2cy - y^2)}.$$

29. Determine the curve in which the radius of curvature is double the normal, and in the opposite direction.

The parabola $x^2 = 4c(y - c)$.

30. Show that the equation

$$\frac{d^2y}{dx^2} + P\frac{dy}{dx} + Q\left(\frac{dy}{dx}\right)^2 = 0$$

can be solved in the following cases: (α) when P and Q are functions of x; (β) when P and Q are functions of y; (γ) when P is a function of x and Q a function of y.

In the case (α), the equation is of the "extended linear form," Art. 37, for $\frac{dy}{dx}$; in the case (β), x does not occur, as in Art. 88; and in the case (γ), the equation is exact when divided by $\frac{dy}{dx}$.

In the last case, the equation may also be solved by the method of variation of parameters, the assumed form of $\frac{dy}{dx}$ being derived by neglecting the last term; the result is

$$\int e^{\int Q dy} dy = A \int e^{-\int P dx} dx + B.$$

CHAPTER V.

LINEAR EQUATIONS WITH CONSTANT COEFFICIENTS.

IX.

Properties of the Linear Equation.

92. A linear differential equation is an equation of the first degree with respect to y and its derivatives. The linear equation of the nth order may therefore be written in the form

$$P_0 \frac{d^n y}{dx^n} + P_1 \frac{d^{n-1} y}{dx^{n-1}} + \ldots + P_n y = X, \quad \ldots \quad (1)$$

in which the coefficients $P_0, P_1, \ldots P_n$ may either be constants or functions of x, and the second member X is generally a function of x.

We have occasion to consider solutions of linear equations only in the form $y = f(x)$, and it is convenient to call a value of y in terms of x which satisfies the equation an *integral* of the equation. Thus, if y_1 is a function of x, such that $y = y_1$ satisfies equation (1), we shall speak of the function y_1, rather than of the equation $y = y_1$, as an integral of equation (1).

93. The solution of equation (1), whether the coefficients be variable or constant, is intimately connected with that of

$$P_0 \frac{d^n y}{dx^n} + P_1 \frac{d^{n-1} y}{dx^{n-1}} + \ldots + P_n y = 0, \quad \ldots \quad (2)$$

which differs from it only in having zero for its second member.

Let y_1 be an integral of equation (2); then C_1y_1, where C_1 is an arbitrary constant, is also an integral. For, if we put $y = C_1y_1$ in the first member, the result is the product by C_1 of the result of substituting $y = y_1$; and, since the latter result vanishes, the former will also vanish.

Again, let y_2 be another integral of equation (2), which is not of the form C_1y_1; then will C_2y_2 be an integral, and $C_1y_1 + C_2y_2$ will also be an integral. For the result of putting $y = C_1y_1 + C_2y_2$ in the first member will be the sum of the results of putting $y = C_1y_1$ and $y = C_2y_2$ respectively, and will therefore vanish. In like manner, if $y_1, y_2, y_3 \ldots y_n$ are n distinct integrals of equation (2),

$$y = C_1y_1 + C_2y_2 + \ldots + C_ny_n \quad \ldots \quad (3)$$

will satisfy the equation; and, since this expression contains n arbitrary constants, it will be the complete integral of equation (2). Thus the complete integral is known when n particular integrals are known, provided they are distinct; that is to say, such that no one can be expressed as a sum of multiples of the others.

94. Now let Y denote a particular integral of the more general equation (1), and let u denote the second member of equation (3), that is to say, the complete integral of equation (2). If we substitute

$$y = Y + u \quad \ldots \ldots \ldots (4)$$

in the first member of equation (1), the result will be the sum of the results of putting $y = Y$, and $y = u$ respectively. The first of these results will be X because Y satisfies equation (1), the second result will be zero because u satisfies equation (2); hence the entire result will be X, and equation (4) is an integral of equation (1). Moreover, it is the complete integral because u contains n arbitrary constants. Thus the complete integral

§ IX.] *PROPERTIES OF THE LINEAR EQUATION.* 93

of equation (1) is known when any one particular integral is known, together with the complete integral of equation (2).

In equation (4), Y is called *the particular integral*, and u is called *the complementary function*. The particular integral contains no arbitrary constants, and any two particular integrals may differ by any multiples of one or more terms belonging to the complementary function.

Linear Equations with Constant Coefficients and Second Member Zero.

95. In the equation

$$A_0 \frac{d^n y}{dx^n} + A_1 \frac{d^{n-1} y}{dx^{n-1}} + \ldots + A_{n-1} \frac{dy}{dx} + A_n y = 0, \quad . \quad (1)$$

in which the coefficients $A_0, A_1 \ldots A_n$ are constants, let us substitute $y = e^{mx}$ where m is a constant to be determined. Since $\frac{d}{dx} e^{mx} = m e^{mx}$, $\frac{d^2}{dx^2} e^{mx} = m^2 e^{mx}$, etc.; the result, after rejecting the factor e^{mx}, is

$$A_0 m^n + A_1 m^{n-1} + \ldots + A_{n-1} m + A_n = 0, \quad . \quad . \quad (2)$$

an equation of the nth degree to determine m. Hence, if m satisfies equation (2), e^{mx} is an integral of equation (1); and, if $m_1, m_2 \ldots m_n$ are n distinct roots of equation (2),

$$y = C_1 e^{m_1 x} + C_2 e^{m_2 x} + \ldots + C_n e^{m_n x} \quad . \quad . \quad . \quad (3)$$

is, by Art. 93, the complete integral of equation (1).

For example, let the given equation be

$$\frac{d^2 y}{dx^2} - \frac{dy}{dx} - 2y = 0;$$

the equation to determine m is

$$m^2 - m - 2 = 0,$$

whose roots are -1 and 2; therefore the complete integral is

$$y = C_1 e^{-x} + C_2 e^{2x}.$$

96. Denoting the symbol $\dfrac{d}{dx}$ by D, equation (1) of Art. 95 may be written

$$(A_0 D^n + A_1 D^{n-1} + \ldots + A_{n-1} D + A_n) y = 0,$$

or, symbolically,

$$f(D) y = 0, \quad \ldots \ldots \ldots \quad (1)$$

in which f denotes a rational integral function. With this notation, equation (2) of the preceding article becomes

$$f(m) = 0;$$

and, denoting its roots, as before, by $m_1, m_2 \ldots m_n$, equation (1) may, in accordance with the principles of commutative and distributive operations (Diff. Calc., Art. 406 *et seq.*), be written in the form

$$(D - m_1)(D - m_2) \ldots (D - m_n) y = 0. \quad \ldots \quad (2)$$

This form of the equation shows that it is satisfied by each of the values of y which separately satisfy the equations

$$(D - m_1) y = 0, \quad (D - m_2) y = 0, \quad \ldots \quad (D - m_n) y = 0;$$

that is to say, by each of the terms of the complete integral.

Thus the example given in the preceding article may be written

$$(D+1)(D-2)y = 0,$$

and the separate terms of the complete integral are the integrals of

$$(D+1)y = 0 \quad \text{and} \quad (D-2)y = 0,$$

which are $C_1 e^{-x}$ and $C_2 e^{2x}$ respectively.

Case of Equal Roots.

97. When two or more roots of the equation $f(m) = 0$ are equal, the general solution, equation (3), Art. 95, fails to represent the complete integral; for, if $m_1 = m_2$, the corresponding terms reduce to

$$(C_1 + C_2)e^{m_1 x},$$

in which $C_1 + C_2$ is equivalent to a single arbitrary constant. It is necessary then to obtain another particular integral; namely, a particular integral of

$$(D - m_1)^2 y = 0, \quad \ldots \quad \ldots \quad (1)$$

in addition to that which also satisfies $(D - m_1)y = 0$.

This integral is obviously the solution of

$$(D - m_1)y = A e^{m_1 x}; \quad \ldots \quad \ldots \quad (2)$$

for, if we apply the operation $D - m_1$ to both members of this equation, we obtain equation (1). Equation (2) is a linear equation of the first order, and its complete integral is

$$e^{-m_1 x} y = \int A dx = Ax + B,$$

or
$$y = e^{m_1 x}(Ax + B). \quad \ldots \ldots \quad (3)$$

Hence the terms of the integral of $f(D)y = 0$ corresponding to a double root of $f(m) = 0$ are found by replacing the constant of integration by $Ax + B$. For example, given the equation
$$\frac{d^3y}{dx^3} - 2\frac{d^2y}{dx^2} + \frac{dy}{dx} = 0,$$
or
$$D(D-1)^2 y = 0,$$
the roots of $f(m) = 0$ are 0, 1, 1, and the complete integral is
$$y = C + e^x(Ax + B).$$

98. If there be three roots equal to m_1, we have, in like manner, to solve
$$(D - m_1)^3 y = 0. \quad \ldots \ldots \quad (1)$$
But the integral of this is the same as that of
$$(D - m_1)y = e^{m_1 x}(Ax + B); \quad \ldots \ldots \quad (2)$$
for, by the preceding article, if the operation $(D - m_1)^2$ be applied to each member of this equation, the result will be $(D - m_1)^3 y = 0$. The integral of equation (2) is
$$e^{-m_1 x} y = \int (Ax + B)dx = \tfrac{1}{2}Ax^2 + Bx + C;$$
or, writing A in place of $\tfrac{1}{2}A$,
$$y = e^{m_1 x}(Ax^2 + Bx + C). \quad \ldots \ldots \quad (3)$$

Hence the terms corresponding to a triple root of $f(m) = 0$ are found by replacing the constant of integration by the expression $Ax^2 + Bx + C$. In like manner, we may show that the terms corresponding to an r-fold root m_1 are

$$e^{m_1 x}(Ax^{r-1} + Bx^{r-2} + \ldots + L).$$

In particular, if the r-fold root is zero, we have for the integral of

$$\frac{d^r y}{dx^r} = 0,$$

$$y = Ax^{r-1} + Bx^{r-2} + \ldots + L,$$

as immediately verified by successive integration.

Case of Imaginary Roots.

99. When the equation $f(m) = 0$ has a pair of imaginary roots, the corresponding terms in the complete integral, as given by the general expression, take an imaginary form; but, assuming the corresponding constants of integration to be also imaginary, the pair of terms is readily reduced to a real form. Thus, if $m_1 = a + i\beta$ and $m_2 = a - i\beta$, the terms in question are

$$C_1 e^{(a+i\beta)x} + C_2 e^{(a-i\beta)x} = e^{ax}(C_1 e^{i\beta x} + C_2 e^{-i\beta x}) \ldots \quad (1)$$

Separating the real and imaginary parts of $e^{i\beta x}$ and $e^{-i\beta x}$, the expression becomes

$$e^{ax}[(C_1 + C_2)\cos\beta x + i(C_1 - C_2)\sin\beta x];$$

or, putting $C_1 + C_2 = A$ and $i(C_1 - C_2) = B$,

$$e^{ax}(A\cos\beta x + B\sin\beta x), \quad \ldots \ldots \quad (2)$$

where, in order that A and B may be real, C_1 and C_2 in (1) must be assumed imaginary.

As an example, let the given equation be

$$(D^2 + D + 1)y = 0;$$

the roots are $-\frac{1}{2} \pm \frac{1}{2}i\sqrt{3}$; here $\alpha = -\frac{1}{2}$, $\beta = \frac{1}{2}\sqrt{3}$; hence the complete integral is

$$y = e^{-\frac{1}{2}x}\left(A \cos\frac{\sqrt{3}}{2}x + B \sin\frac{\sqrt{3}}{2}x\right).$$

100. If the equation $f(m) = 0$ has a pair of imaginary r-fold roots, we must, by Art. 98, replace each of the arbitrary constants in expression (1) by a polynomial of the $(r-1)$th degree; whence it readily follows that we must, in like manner, replace the constants in expression (2) by similar polynomials. Thus the equation

$$\frac{d^4y}{dx^4} + 2\frac{d^2y}{dx^2} + y = 0,$$

or

$$(D^2 + 1)^2 y = 0,$$

in which $\pm i$ are double roots, has for its integral

$$y = (A_1 + B_1 x)\cos x + (A_2 + B_2 x)\sin x.$$

The Linear Equation with Constant Coefficients and Second Member a Function of x.

101. In accordance with the symbolic notation, the value of y which satisfies the equation

$$f(D)y = X \quad \dots \dots \quad (1)$$

is denoted by

$$y = \frac{1}{f(D)} X. \quad \dots \dots \quad (2)$$

Substituting this expression in equation (1), we have

$$f(D)\frac{1}{f(D)}X = X,$$

which may be regarded as defining the inverse symbol (2), so that it denotes *any* function of X which, when operated upon by the direct symbol $f(D)$, produces the given function X. Then, by Art. 94, the complete integral of equation (1) is the sum of any legitimate value of the inverse symbol and the complementary function or complete integral of

$$f(D)y = 0.$$

This last function, which is found by the methods explained in the preceding articles, we may call the complementary function for $f(D)$; and we see that two legitimate values of the symbol $\frac{1}{f(D)}X$ may differ by an arbitrary multiple of any term in the complementary function for $f(D)$; just as two values of $\int X dx$ or $\frac{1}{D}X$ may differ by an arbitrary constant, which is the complementary function for D.

102. With this understanding of the indefinite character of the inverse symbols, it is evident that an equation involving such symbols is admissible, provided only it is reducible to an identity by performing the necessary direct operations upon each member. It follows that the inverse symbols may be transformed exactly as if they represented algebraic quantities; for, owing to the commutative and distributive character of the direct operations, the process of verifying the equation is precisely the same whether it be regarded as symbolic or algebraic. For example, to verify the symbolic identity

$$\frac{1}{D^2 - a^2}X = \frac{1}{2a}\left(\frac{1}{D-a}X - \frac{1}{D+a}X\right),$$

we perform the operation $D^2 - a^2$ on both members; thus

$$X = \frac{1}{2a}\left[(D+a)(D-a)\frac{1}{D-a}X - (D-a)(D+a)\frac{1}{D+a}X\right]$$
$$= \frac{1}{2a}\left[(D+a)X - (D-a)X\right] = \frac{1}{2a}2aX = X,$$

the process being equivalent to that of verifying the equation

$$\frac{1}{D^2 - a^2} = \frac{1}{2a}\left(\frac{1}{D-a} - \frac{1}{D+a}\right)$$

considered as an algebraic identity.

103. The symbol $\dfrac{1}{D-a}X$ denotes the value of y in the equation of the first order

$$\frac{dy}{dx} - ay = X;$$

hence, solving, we have

$$\frac{1}{D-a}X = e^{ax}\int e^{-ax}X dx.. \quad \ldots \quad (1)$$

By repeated application of this formula, we have

$$\frac{1}{(D-a)^2}X = \frac{1}{D-a}e^{ax}\int e^{-ax}X dx = e^{ax}\int\int e^{-ax}X dx dx; \quad (2)$$

and, in general,

$$\frac{1}{(D-a)^r}X = e^{ax}\int\int\ldots\int e^{-ax}X dx^r, \quad \ldots \quad (3)$$

the last expression involving an integral of the rth order.

General Expression for the Integral.

104. We may, by means of equation (1) of the preceding article, write an expression for the complete integral of $f(D)y = X$ involving a multiple integral of the nth order. For, using the notation of preceding articles, we may put

$$f(D) = (D - m_1)(D - m_2) \ldots (D - m_n);$$

whence

$$\frac{1}{f(D)}X = \frac{1}{D - m_1}\frac{1}{D - m_2}\cdots\frac{1}{D - m_n}X$$

$$= e^{m_1 x}\int e^{(m_2 - m_1)x}\int \ldots \int e^{-m_n x}X dx^n; \quad \ldots \quad (1)$$

but the expression given below is preferable, involving, as it does, multiple integrals only when the equation $f(D) = 0$ has multiple roots.

105. Let $\dfrac{1}{f(D)}$ be resolved into partial fractions; supposing $m_1, m_2 \ldots m_n$ to be all different, the result will be of the form

$$\frac{1}{f(D)} = \frac{N_1}{D - m_1} + \frac{N_2}{D - m_2} + \ldots + \frac{N_n}{D - m_n}, \quad \cdot \quad (1)$$

in which $N_1, N_2 \ldots N_n$ are determinate constants; hence, by equation (1), Art. 103,

$$\frac{1}{f(D)}X = N_1 e^{m_1 x}\int e^{-m_1 x}X dx + \ldots + N_n e^{m_n x}\int e^{-m_n x}X dx, \quad (2)$$

which is the general expression[*] for the complete integral

---

[*] First published by Lobatto, "Théorie des Caractéristiques," Amsterdam, 1837; independently discovered by Boole, *Cambridge Math. Journal*, 1st series. vol. ii. p. 114.

when the roots of $f(D) = 0$ are all different; each term, it will be noticed, containing one term of the complementary function.

When two of the roots of $f(D) = 0$ are equal, say $m_1 = m_2$, the corresponding partial fractions in equation (1) must be assumed in the form

$$\frac{N_1}{D - m_1} + \frac{N_2}{(D - m_1)^2};$$

and then by equations (1) and (2), Art. 103, the corresponding terms in equation (2) will be

$$N_1 e^{m_1 x} \int e^{-m_1 x} X dx + N_2 e^{m_1 x} \int\int e^{-m_1 x} X dx dx.$$

In like manner, a multiple root of the rth order gives rise to multiple integrals of the rth and lower orders.

106. When $f(D) = 0$ has a pair of imaginary roots, $a \pm i\beta$, we may first determine, for the corresponding quadratic factor, a partial fraction of the form

$$\frac{N_1 D + N_2}{(D - a)^2 + \beta^2}.$$

The corresponding part of the integral will be found by applying the operation $N_1 D + N_2$ to the value of

$$\frac{1}{(D - a)^2 + \beta^2} X.$$

Decomposing the symbolic operator further, this expression becomes

$$\frac{1}{2i\beta} \left(\frac{1}{D - a - i\beta} - \frac{1}{D - a + i\beta} \right) X;$$

that is,

$$\frac{1}{2i\beta} e^{(a + i\beta)x} \int e^{-(a + i\beta)x} X dx - \frac{1}{2i\beta} e^{(a - i\beta)x} \int e^{-(a - i\beta)x} X dx.$$

This last expression is the sum of two terms of which the second is the same as the first with the sign of i changed; and, the first term being a complex quantity of the form $P + iQ$ where P and Q are real, the sum is $2P$, or twice the real part of the first term. Hence

$$\frac{1}{(D-a)^2 + \beta^2} X$$

$$= \text{the real part of } \frac{e^{ax}}{i\beta}(\cos\beta x + i\sin\beta x)\int e^{-ax}(\cos\beta x - i\sin\beta x)Xdx,$$

or

$$\frac{1}{(D-a)^2 + \beta^2} X$$

$$= \frac{e^{ax}\sin\beta x}{\beta}\int e^{-ax}\cos\beta x \, Xdx - \frac{e^{ax}\cos\beta x}{\beta}\int e^{-ax}\sin\beta x \, Xdx.$$

When $a = 0$, this result reduces to that otherwise found in Arts. 91 and 82.

Examples IX.

Solve the following differential equations:—

1. $\dfrac{d^2y}{dx^2} - 5\dfrac{dy}{dx} + 6y = 0,$ $\qquad y = c_1 e^{2x} + c_2 e^{3x}.$

2. $6\dfrac{d^2y}{dx^2} = \dfrac{dy}{dx} + y,$ $\qquad y = c_1 e^{\frac{1}{2}x} + c_2 e^{-\frac{1}{3}x}.$

3. $ab\left(\dfrac{d^2y}{dx^2} + y\right) = (a^2 + b^2)\dfrac{dy}{dx},$ $\qquad y = c_1 e^{\frac{ax}{b}} + c_2 e^{\frac{bx}{a}}.$

4. $\dfrac{d^2y}{dx^2} - 2\dfrac{dy}{dx} + 5y = 0,$ $y = e^x(A\cos 2x + B\sin 2x).$

5. $4\dfrac{d^4y}{dx^4} - 3\dfrac{d^2y}{dx^2} - y = 0,$

$$y = c_1 e^x + c_2 e^{-x} + A\sin\tfrac{1}{2}(x + a).$$

6. $\dfrac{d^3y}{dx^3} + \dfrac{d^2y}{dx^2} - 6\dfrac{dy}{dx} = 0,$ $y = c_1 e^{2x} + c_2 e^{-3x} + c_3.$

7. $\dfrac{d^3y}{dx^3} - 7\dfrac{dy}{dx} + 6y = 0,$ $y = c_1 e^{2x} + c_2 e^{-3x} + c_3 e^x.$

8. $4\dfrac{d^3y}{dx^3} - 3\dfrac{dy}{dx} + y = 0,$ $y = e^{\frac{1}{2}x}(Ax + B) + ce^{-x}.$

9. $\dfrac{d^4y}{dx^4} + 4y = 0,$

$$y = (c_1 e^x + c_2 e^{-x})\cos x + (c_3 e^x + c_4 e^{-x})\sin x.$$

10. $\dfrac{d^3y}{dx^3} - \dfrac{d^2y}{dx^2} - \dfrac{dy}{dx} + y = 0,$ $y = c_1 e^{-x} + (c_2 + c_3 x)e^x.$

11. $\dfrac{d^4y}{dx^4} + 2\dfrac{d^3y}{dx^3} = 2\dfrac{dy}{dx} + y,$

$$y = c_1 e^x + e^{-x}(c_2 + c_3 x + c_4 x^2).$$

12. $\dfrac{d^4y}{dx^4} - 4\dfrac{d^3y}{dx^3} + 8\dfrac{d^2y}{dx^2} - 8\dfrac{dy}{dx} + 4y = 0,$

$$y = e^x(c_1 + c_2 x)\sin x + e^x(c_3 + c_4 x)\cos x.$$

13. $\dfrac{d^2y}{dx^2} + a^2 y = \sec ax,$

$$y = c_1 \cos ax + c_2 \sin ax + \dfrac{x\sin ax}{a} + \dfrac{\cos ax \log \cos ax}{a^2}.$$

14. $\dfrac{d^2y}{dx^2} + y = \sec^2 x$,

$$y = A\cos(x+a) + \sin x \log\dfrac{1+\sin x}{\cos x} - 1.$$

15. $\dfrac{d^2y}{dx^2} + y = \tan x$, $\quad y = A\cos(x+a) - \cos x \log\dfrac{1+\sin x}{\cos x}.$

16. Show that $\dfrac{D}{(D-a)^2 + \beta^2}X$

$$= \dfrac{e^{ax}}{\beta}\left[(a\sin\beta x + \beta\cos\beta x)\int e^{-ax}\cos\beta x X dx\right.$$
$$\left. - (a\cos\beta x - \beta\sin\beta x)\int e^{-ax}\sin\beta x X dx\right].$$

17. Show that $\dfrac{1}{(D^2+a^2)^2}X$

$$= \dfrac{1}{2a^3}\left[\sin ax\int\cos ax X dx - \cos ax\int\sin ax X dx\right]$$
$$- \dfrac{1}{2a^2}\left[\cos ax\iint\cos ax X dx^2 + \sin ax\iint\sin ax X dx^2\right].$$

18. $\dfrac{d^3y}{dx^3} - 2\dfrac{dy}{dx} + 4y = X$,

$$y = c_1 e^{-2x} + e^x(c_2\cos x + c_3\sin x) + \dfrac{e^{-2x}}{10}\int e^{2x}X dx$$
$$+ \dfrac{e^x}{10}\left[(3\sin x - \cos x)\int e^{-x}\cos x X dx\right.$$
$$\left. - (3\cos x + \sin x)\int e^{-x}\sin x X dx\right].$$

X.

Symbolic Methods of Integration.

107. The foregoing general solution of linear equations with constant coefficients, Art. 105, is theoretically complete; for the solution of a differential equation consists in finding a relation between x and y involving only the integral sign. But, in the case of certain forms of the function X of frequent occurrence, while the evaluation of the integrals arising in the general solution would be tedious, the final result may be very expeditiously obtained by the methods now to be explained.

In the first place, *suppose the second member X to be of the form e^{ax}*; in other words, let it be required to solve the equation

$$f(D)y = e^{ax}. \qquad (1)$$

Since, as in Art. 95, $D^r e^{ax} = a^r e^{ax}$, and $f(D)$ is a sum of terms of the form AD^r,

$$f(D)e^{ax} = f(a)e^{ax}; \qquad (2)$$

whence

$$\frac{1}{f(D)}f(a)e^{ax} = e^{ax}.$$

Here $f(a)$ is a constant; and therefore, except when $f(a) = 0$, we may divide by it and write

$$\frac{1}{f(D)}e^{ax} = \frac{1}{f(a)}e^{ax}, \qquad (3)$$

which is the value of y in equation (1). Thus we may, when the operand is of the form Ae^{ax}, put $D = a$ in the operating symbol except when the result would introduce an infinite coefficient.

108. In the exceptional case, equation (2), of course, still holds; but it reduces to $f(D)e^{ax} = 0$, and thus only expresses that e^{ax} is a term of the complementary function. In this case, we may still put a for D in all the factors of $f(D)$ except $D - a$. Thus, putting

$$f(D) = (D - a)\phi(D),$$

we have

$$\frac{1}{f(D)}e^{ax} = \frac{1}{D-a}\frac{1}{\phi(D)}e^{ax} = \frac{1}{\phi(a)}\frac{1}{D-a}e^{ax};$$

and hence, by equation (1), Art. 103,

$$\frac{1}{f(D)}e^{ax} = \frac{1}{\phi(a)}e^{ax}\int e^{-ax}e^{ax}dx = \frac{xe^{ax}}{\phi(a)}.$$

Again, if $f(D) = (D - a)^2\phi(D)$, so that a is a double root of $f(D) = 0$, we shall have

$$\frac{1}{f(D)}e^{ax} = \frac{1}{(D-a)^2}\frac{1}{\phi(D)}e^{ax} = \frac{1}{\phi(a)}e^{ax}\iint dx^2 = \frac{x^2 e^{ax}}{2\phi(a)}.$$

109. As an illustration, let it be required to solve the equation

$$\frac{d^3y}{dx^3} - y = (e^x + 1)^2. \quad \ldots \ldots (1)$$

The complementary function is

$$Ce^x + e^{-\frac{1}{2}x}\left(A\cos\frac{\sqrt{3}}{2}x + B\sin\frac{\sqrt{3}}{2}x\right).$$

The particular integral is

$$y = \frac{1}{D^3 - 1}(e^x + 1)^2 = \frac{1}{D^3 - 1}e^{2x} + 2\frac{1}{D^3 - 1}e^x + \frac{1}{D^3 - 1}e^0.$$

In the first and third terms, we may put $D = 2$ and 0

respectively, thus obtaining $\frac{1}{3}e^{ax} - 1$; but $D = 1$ makes the second term infinite; hence we write

$$2\frac{1}{D^3 - 1}e^x = 2\frac{1}{D - 1}\frac{1}{D^2 + D + 1}e^x$$

$$= \tfrac{2}{3}\frac{1}{D - 1}e^x = \tfrac{2}{3}e^x\!\int\! dx = \tfrac{2}{3}xe^x.$$

The complete integral of equation (1) is therefore

$$y = e^{-\frac{1}{2}x}\!\left(A\cos\tfrac{\sqrt{3}}{2}x + B\sin\tfrac{\sqrt{3}}{2}x\right) + e^x(C + \tfrac{2}{3}x) + \tfrac{1}{3}e^{ax} - 1.$$

110. The value of the particular integral in the case of failure of equation (3), Art. 107, may also be derived directly from that equation by the principle of continuity. It must be remembered that properly the equation should be understood to contain the complementary function in the second member. Hence, a being a root of $f(D) = 0$, and at first assuming the operand to be $e^{(a+h)x}$, we may write

$$\frac{1}{f(D)}e^{(a+h)x} = \frac{1}{f(a+h)}e^{ax}e^{hx} + Ce^{ax} + \cdots.$$

Developing e^{hx}, the second member becomes

$$\left[\frac{1}{f(a+h)} + C\right]e^{ax} + \frac{e^{ax}}{f(a+h)}\!\left(hx + \frac{h^2x^2}{2} + \cdots\right) + \cdots,$$

in which the first term is part of the complementary function. We may therefore write, for the particular integral,

$$\frac{1}{f(D)}e^{(a+h)x} = \frac{h}{f(a+h)}xe^{ax}\!\left(1 + \frac{hx}{2} + \cdots\right)$$

$$= \frac{1}{\phi(a+h)}xe^{ax}\!\left(1 + \frac{hx}{2} + \cdots\right)$$

because, a being a root of $f(z) = 0$, $f(z) = (z - a)\phi(z)$, and $f(a + h) = h\phi(a + h)$.

§ X.] SYMBOLIC METHODS OF INTEGRATION. 109

Now, making $h = 0$ in this result, we obtain

$$\frac{1}{f(D)} e^{ax} = \frac{1}{\phi(a)} x e^{ax},$$

as before.

This is an instance of a general principle of which we shall hereafter meet other applications; namely, that, when the particular integral, as given by a general formula, becomes infinite, it can be developed into an infinite term which merges into the complementary function, and a finite part which furnishes a new particular integral.

Again, when a is a double root, and $X = e^{ax}$, the infinite expression can be developed into two infinite terms which merge into the complementary function, together with a finite term which gives the new particular integral. For example, since h is ultimately to be put equal to zero, we may write

$$\frac{1}{(D-a)^2 \phi(D)} e^{ax} = \frac{1}{\phi(a+h) h^2} e^{(a+h)x}$$

$$= \frac{e^{ax}}{\phi(a+h) h^2} \left(1 + hx + \frac{h^2 x^2}{2} + \ldots\right).$$

The first two terms have infinite coefficients when $h = 0$, but they belong to the complementary function; the third term is finite, and gives the particular integral

$$\frac{1}{(D-a)^2 \phi(D)} e^{ax} = \frac{x^2 e^{ax}}{2\phi(a)}.$$

Case in which X contains a Term of the Form $\sin ax$ or $\cos ax$.

111. We have, by differentiation,

$$D \sin ax = a \cos ax, \quad D^2 \sin ax = -a^2 \sin ax,$$
$$D^{2r} \sin ax = (-a^2)^r \sin ax;$$

whence

$$f(D^2) \sin ax = f(-a^2) \sin ax.$$

and, in like manner, we obtain

$$f(D^2)\cos ax = f(-a^2)\cos ax.$$

It follows, as in the similar case of Art. 107, that

$$\frac{1}{f(D^2)}\sin ax = \frac{1}{f(-a^2)}\sin ax, \ldots \ldots (1)$$

and

$$\frac{1}{f(D^2)}\cos ax = \frac{1}{f(-a^2)}\cos ax, \ldots \ldots (2)$$

except when $f(-a^2) = 0$. It is obvious that we may include both these results in the slightly more general formula

$$\frac{1}{f(D^2)}\sin(ax+\alpha) = \frac{1}{f(-a^2)}\sin(ax+\alpha).$$

For example, to solve

$$\frac{d^2y}{dx^2} - y = \sin(x+\alpha),$$

we have, for the particular integral,

$$\frac{1}{D^2 - 1}\sin(x+\alpha) = -\tfrac{1}{2}\sin(x+\alpha).$$

Adding the complementary function, we have the complete integral

$$y = c_1 e^x + c_2 e^{-x} - \tfrac{1}{2}\sin(x+\alpha).$$

112. In order to employ equations (1) and (2) when the inverse symbol is not a function of D^2, we reduce it to a fractional form in which the denominator is a function of

§ X.] SECOND MEMBER OF THE FORM $\sin ax$ OR $\cos ax$.

D^2. This is readily done; for we may put $f(D)$ in the form $f_1(D^2) + Df_2(D^2)$, and the product of this by $f_1(D^2) - Df_2(D^2)$ will be a function of D^2. Moreover, since we have ultimately to put $D^2 = -a^2$, we may at once put $-a^2$ in place of D^2 in the expression for $f(D)$, which thus becomes

$$f_1(-a^2) + Df_2(-a^2).$$

For example, given the equation

$$(D^2 + D - 2)y = \sin 2x;$$

the particular integral is

$$\frac{1}{D^2 + D - 2}\sin 2x = \frac{1}{D - 6}\sin 2x = \frac{D + 6}{D^2 - 36}\sin 2x$$

$$= -\frac{D + 6}{40}\sin 2x = -\frac{\cos 2x + 3\sin 2x}{20}.$$

Adding the complementary function

$$y = C_1 e^x + C_2 e^{-2x} - \frac{\cos 2x + 3\sin 2x}{20}.$$

113. The case of failure of the formulæ (1) and (2) of Art. 111 takes place when the operand is a term of the complementary function. Thus, if the given equation is

$$\frac{d^2y}{dx^2} + a^2 y = \cos ax,$$

the complementary function is $A \cos ax + B \sin ax$. Accordingly, in the particular integral $\dfrac{1}{D^2 + a^2} \cos ax$, the substitution $D^2 = -a^2$ gives an infinite coefficient. The most convenient method of evaluating in this case is that illustrated in Art. 110. Thus, putting $a + h$ for a in the operand, and developing $\cos(ax + hx)$ by Taylor's theorem,

$$\frac{1}{D^2 + a^2}\cos(a + h)x$$

$$= \frac{1}{-(a+h)^2 + a^2}\left(\cos ax - \sin ax \cdot hx - \cos ax \cdot \frac{h^2 x^2}{2} + \ldots\right).$$

Omitting the first term which belongs to the complementary function, we may write, for the particular integral,

$$\frac{1}{D^2 + a^2}\cos(a + h)x = \frac{1}{2a + h}\left(x\sin ax + \frac{hx^2}{2}\cos ax - \ldots\right);$$

and, making $h = 0$, we obtain

$$\frac{1}{D^2 + a^2}\cos ax = \frac{x\sin ax}{2a},$$

and the complete integral of equation (1) is

$$y = A\cos ax + B\cos ax + \frac{x\sin ax}{2a}.$$

Case in which X contains Terms of the Form x^m.

114. If an inverse symbol be developed into a series proceeding by ascending powers of D, the result of operating upon a function of x with the transformed symbol is, in general, an infinite series of functions; but, when the operand is of the form x^m, where m is a positive integer, the derivatives above the mth vanish, and the result is finite. For example, to solve

$$\frac{dy}{dx} + 2y = x^3,$$

the particular integral is

$$\frac{1}{D+2}x^3 = \frac{1}{2}\frac{1}{1+\tfrac{1}{2}D}x^3$$
$$= \tfrac{1}{2}(1 - \tfrac{1}{2}D + \tfrac{1}{4}D^2 - \tfrac{1}{8}D^3 + \ldots)x^3$$
$$= \tfrac{1}{2}(x^3 - \tfrac{3}{2}x^2 + \tfrac{3}{2}x - \tfrac{3}{4});$$

and the complete integral is

$$y = Ce^{-2x} + \tfrac{1}{2}x^3 - \tfrac{3}{4}x^2 + \tfrac{3}{4}x - \tfrac{3}{8}.$$

This result is readily verified by performing upon it the operation $D + 2$.

115. When the denominator of the inverse symbol is divisible by a power of D, the development will commence with a negative power of D, but no greater number of terms will be required than would be were the factor D not present. For example, if the given equation is

$$(D^4 + D^3 + D^2)y = x^3 + 3x^2,$$

the particular integral is

$$y = \frac{1}{D^4 + D^3 + D^2}(x^3 + 3x^2) = \frac{1}{D^2}\frac{1}{1 + D + D^2}(x^3 + 3x^2)$$
$$= \frac{1}{D^2}\Big[1 - (D + D^2) + (D + D^2)^2 - (D + D^2)^3 + \ldots\Big](x^3 + 3x^2).$$

Since the operand contains no power of x higher than x^3, it is unnecessary to retain powers of D higher than D^3 in the development of the expression in brackets. Hence we write

$$y = \frac{1}{D^2}(1 - D + D^3)(x^3 + 3x^2) = \Big(\frac{1}{D^2} - \frac{1}{D} + D\Big)(x^3 + 3x^2)$$
$$= \frac{x^5}{20} + \frac{x^4}{4} - \frac{x^4}{4} - x^3 + 3x^2 + 6x,$$

in which the last term should be rejected as included in the complementary function. Thus the complete integral is

$$y = \frac{x^5}{20} - x^3 + 3x^2 + c_1 x + c_2 + e^{-\frac{1}{2}x}\left(c_3 \cos\frac{\sqrt{3}}{2}x + c_4 \sin\frac{\sqrt{3}}{2}x\right).$$

It will be noticed that, had we retained any higher powers of D in the final development, they would have produced only terms included in the complementary function.

Symbolic Formulæ of Reduction.

116. The formulæ of reduction explained in this and the following articles apply to cases in which X contains a factor of a special form.

In the first place, *let X be of the form $e^{ax}V$, V being any function of x.* By differentiation,

$$\frac{d}{dx}e^{ax}V = e^{ax}\frac{dV}{dx} + ae^{ax}V,$$

or

$$De^{ax}V = e^{ax}(D+a)V. \quad \ldots \ldots \quad (1)$$

By repeated application of this formula, we have

$$D^2 e^{ax}V = De^{ax}(D+a)V = e^{ax}(D+a)^2 V;$$

and, in general,

$$D^r e^{ax}V = e^{ax}(D+a)^r V.$$

Hence, when $\phi(D)$ is a direct symbol involving integral powers of D, we have

$$\phi(D)e^{ax}V = e^{ax}\phi(D+a)V. \quad \ldots \ldots \quad (2)$$

§ X.] SYMBOLIC FORMULÆ OF REDUCTION.

To show that this formula is applicable also to inverse symbols, put
$$\phi(D + a)V = V_1;$$
whence
$$V = \frac{1}{\phi(D + a)}V_1;$$
and equation (2) becomes
$$e^{ax}V_1 = \phi(D)e^{ax}\frac{1}{\phi(D + a)}V_1,$$
in which V_1 denotes any function of x; since V was unrestricted. Now, applying the operation $\frac{1}{\phi(D)}$ to both members, we have
$$\frac{1}{\phi(D)}e^{ax}V = e^{ax}\frac{1}{\phi(D + a)}V, \quad \ldots \quad (3)$$
which is of the same form as equation (2).

As an example of the application of this formula, let the given equation be
$$\frac{d^2y}{dx^2} - y = xe^{ax}.$$

The particular integral is

117. The formula of reduction of the preceding article may often be used with advantage in the evaluation of an ordinary integral. For example, to find $\int e^{mx} \sin nx\,dx$, we have, by the formula,

$$\frac{1}{D} e^{mx} \sin nx = e^{mx} \frac{1}{D+m} \sin nx\,;$$

hence

$$\int e^{mx} \sin nx\,dx = e^{mx} \frac{D-m}{D^2-m^2} \sin nx$$

$$= \frac{e^{mx}}{m^2+n^2}(m-D)\sin nx = \frac{e^{mx}}{m^2+n^2}(m \sin nx - n \cos nx).$$

It may be noticed that equation (1), Art. 103, is a case of the present formula of reduction, for

$$\frac{1}{D-a} X = \frac{1}{D-a} e^{ax} e^{-ax} X\,;$$

hence, applying the formula, we obtain

$$\frac{1}{D-a} X = e^{ax} \frac{1}{D} e^{-ax} X = e^{ax} \int e^{-ax} X\,dx\,;$$

in which we pass from the solution of a differential equation to a simple integration. In the above example, on the other hand, we employed the same formula to reverse the process, the direct solution of the differential equation being, in that case, the simpler process. Compare Int. Calc., Art. 63.

118. Secondly, *let X be of the form xV*. By successive differentiation, we have

$$DxV = xDV + V,$$
$$D^2xV = xD^2V + 2DV,$$
$$D^3xV = xD^3V + 3D^2V\,;$$

and, generally,

$$D^r xV = xD^r V + rD^{r-1}V. \quad \ldots \ldots \quad (1)$$

Now let $\phi(D)$ denote a rational integral function of D, that is, the sum of terms of the form $a_r D^r$; and let us transform each term of $\phi(D)x V$ by means of equation (1). We thus have two sets of terms whose sums are $x\Sigma a_r D^r V$ and $\Sigma a_r r D^{r-1} V$ respectively. The first sum is obviously $x\phi(D)V$; and, since $a_r r D^{r-1}$ is the derivative of $a_r D^r$ considered as a function of D, the second sum constitutes the function $\phi'(D)V$. Hence

$$\phi(D)xV = x\phi(D)V + \phi'(D)V, \quad \ldots \quad (2)$$

where ϕ' is the derivative of the function ϕ.

To show that this formula is true also for inverse symbols, put
$$\phi(D)V = V_1;$$
whence
$$V = \frac{1}{\phi(D)}V_1;$$
and equation (2) becomes

$$\phi(D)x\frac{1}{\phi(D)}V_1 = xV_1 + \phi'(D)\frac{1}{\phi(D)}V_1,$$

or

$$xV_1 = \phi(D)x\frac{1}{\phi(D)}V_1 - \frac{\phi'(D)}{\phi(D)}V_1,$$

in which V_1 denotes any function of x. Hence, applying the operation $\frac{1}{\phi(D)}$ to both members, we have the general formula

$$\frac{1}{\phi(D)}xV = x\frac{1}{\phi(D)}V - \frac{\phi'(D)}{[\phi(D)]^2}V, \quad \ldots \quad (3)$$

which is of the same form as equation (2), because $-\frac{\phi'}{\phi^2}$ is the derivative of the function $\frac{1}{\phi}$.

119. As an example, take the linear equation

$$\frac{dy}{dx} - y = x \sin x.$$

By the formula, the particular integral is

$$\frac{1}{D-1} x \sin x = x \frac{1}{D-1} \sin x - \frac{1}{(D-1)^2} \sin x$$

$$= x \frac{D+1}{D^2-1} \sin x - \frac{D^2 + 2D + 1}{(D^2-1)^2} \sin x;$$

hence

$$y = -\tfrac{1}{2}x(\cos x + \sin x) - \tfrac{1}{2}\cos x + Ce^x.$$

This example is a good illustration of the advantage of the symbolic method, for the general solution would give the integral in the very inconvenient form

$$.y = e^x \int e^{-x} \sin x \, dx + Ce^x;$$

and, in fact, the best way to evaluate the indefinite integral in this expression is by the symbolic method, as in Art. 117.

120. Finally, *let X be of the form $x^r V$.* Putting xV in place of V in formula (2), Art. 118,

$$\phi(D)x^2 V = x\phi(D)xV + \phi'(D)xV;$$

and, reducing by the same formula the expressions $\phi(D)xV$ and $\phi'(D)xV$, this becomes

$$\phi(D)x^2 V = x^2\phi(D)V + 2x\phi'(D)V + \phi''(D)V. \quad . \quad (4)$$

Again, putting xV for V in this formula, and reducing as before, we have

$$\phi(D)x^3 V = x^3\phi(D)V + 3x^2\phi'(D)V + 3x\phi''(D)V + \phi'''(D)V;$$

and by the same process we obtain similar formulæ for $x^4 V$, $x^5 V$, etc., the numerical coefficients introduced being obviously those of the binomial theorem.

As an illustration, let the given equation be

$$\frac{d^2y}{dx^2} + y = x^2 \sin 2x.$$

By formula (4), the particular integral is

$$\frac{1}{D^2+1} x^2 \sin 2x$$

$$= x^2 \frac{1}{D^2+1} \sin 2x + 2x \frac{-2D}{(D^2+1)^2} \sin 2x + \frac{6D^2-2}{(D^2+1)^3} \sin 2x$$

$$= -\frac{x^2}{3} \sin 2x - \frac{8x}{9} \cos 2x + \frac{26}{27} \sin 2x,$$

and the complete integral is

$$y = c_1 \cos x + c_2 \sin x - \frac{9x^2-26}{27} \sin 2x - \frac{8x}{9} \cos 2x.$$

Employment of the Exponential Forms of sin ax *and* cos ax.

121. It is often useful to substitute for a factor of the form sin ax or cos ax its exponential value, and then to reduce the result by means of formula (2) of Art. 116. For example, in solving the equation

$$\frac{d^2y}{dx^2} + y = x^2 \sin x,$$

we have, for the particular integral,

$$y = \frac{1}{D^2+1} x^2 \sin x = \frac{1}{D^2+1} \frac{x^2}{2i}(e^{ix} - e^{-ix});$$

but it is rather more convenient to write, what is easily seen to be the same thing, since $e^{ix} = \cos x + i \sin x$,

$$y = \text{the coefficient of } i \text{ in } \frac{1}{D^2+1} x^2 e^{ix}.$$

Now
$$\frac{1}{D^2+1}x^2e^{ix} = e^{ix}\frac{1}{(D+i)^2+1}x^2 = e^{ix}\frac{1}{D(D+2i)}x^2$$
$$= \frac{e^{ix}}{2i}\frac{1}{D}\left(1 - \frac{D}{2i} + \frac{D^2}{4i^2} - \ldots\right)x^2$$
$$= e^{ix}\left(-\frac{i}{2D} + \frac{1}{4} + \frac{iD}{8}\right)x^2$$
$$= (\cos x + i\sin x)\left(-\frac{ix^3}{6} + \frac{x^2}{4} + \frac{ix}{4}\right);$$

whence, taking the coefficient of i, and adding the complementary function,

$$y = \cos x(A - \tfrac{1}{6}x^3 + \tfrac{1}{4}x) + \sin x(B + \tfrac{1}{4}x^2).*$$

Examples X.

Solve the following differential equations:—

1. $\dfrac{d^2y}{dx^2} - y = xe^{2x} + e^x$,

$$y = c_1e^x + c_2e^{-x} + \frac{e^{2x}}{9}(3x-4) + \frac{xe^x}{2}.$$

---

* This method has an obvious advantage over that of Art. 120 when a high power of x occurs. Moreover, when, as in the present example, the trigonometrical factor is a term of the complementary function, it should always be employed. For it is to be noticed that, in formula (3), Art. 118, while two legitimate values of the symbol in the first member can differ only by multiples of terms in the complementary function of $\phi(D)$, two values of the second member may differ by the product of one of these terms by x. Hence a result obtained by the formula might be erroneous with respect to the coefficient of such a term. In the example of Art. 119, the uncertainty would exist only with respect to a term of the form xe^x, but it is easy to see that no such term can occur in the solution. In the example of Art. 120, a similar uncertainty exists with respect to terms of the form $x^2\sin x$, $x^2\cos x$, $x\sin x$, and $x\cos x$, none of which occur in the solution. In the present example, if solved by the same method, the uncertainty would exist with respect to terms of the same form; and, as such terms do occur in the solution, an error might arise. See *Messenger of Mathematics*, vol. xvi. p. 86.

§ X.] EXAMPLES. 121

2. $\dfrac{dy}{dx} - 2y = x^3 + e^x + \cos 2x$,

$y = ce^{2x} - e^x - \tfrac{1}{8}(4x^3 + 6x^2 + 6x + 3)$
$\qquad\qquad + \tfrac{1}{4}(\sin 2x - \cos 2x)$.

3. $\dfrac{d^2y}{dx^2} - 2\dfrac{dy}{dx} + y = x^2 e^{3x}$,

$y = (A + Bx)e^x + \dfrac{e^{3x}}{8}(2x^2 - 4x + 3)$.

4. $\dfrac{d^4y}{dx^4} + 2\dfrac{d^2y}{dx^2} + y = \sin x$,

$y = (c_1 x + c_2) \sin x + (c_3 x + c_4) \cos x - \tfrac{1}{8}x^2 \sin x$.

5. $\dfrac{d^2y}{dx^2} - 2\dfrac{dy}{dx} + y = xe^x$, $\qquad y = e^x(c_1 + c_2 x + \tfrac{1}{6}x^3)$.

6. $\dfrac{d^2y}{dx^2} + 4y = \sin 3x + e^x + x^2$,

$y = A \cos 2x + B \sin 2x + \tfrac{1}{5}(e^x - \sin 3x) + \tfrac{1}{8}(2x^2 - 1)$.

7. $\dfrac{d^2y}{dx^2} - 2\dfrac{dy}{dx} + 2y = e^x \sin x + \cos x$,

$y = e^x(A \cos x + B \sin x) - \tfrac{1}{2}xe^x \cos x + \tfrac{1}{5}(\cos x - 2 \sin x)$.

8. $\dfrac{d^2y}{dx^2} + y = x \sin 2x$,

$y = A \cos x + B \sin x - \tfrac{1}{3}x \sin 2x - \tfrac{4}{9} \cos 2x$.

9. $\dfrac{d^2y}{dx^2} + y = x \sin x$,

$y = A \cos x + B \sin x - \dfrac{x^2}{4} \cos x + \dfrac{x}{4} \sin x$.

10. $\dfrac{d^2y}{dx^2} + 4y = 2x^3 \sin^2 x$,

$y = A \sin 2x + B \cos 2x + \dfrac{2x^3 - 3x}{8}$

$\qquad - \dfrac{8x^3 - 3x}{128} \cos 2x - \dfrac{4x^4 - 3x^2}{64} \sin 2x$.

11. $\dfrac{d^4y}{dx^4} - y = e^x \cos x$,

$\qquad y =$ Complementary Function $- \tfrac{1}{6} e^x \cos x$.

12. $\dfrac{d^6y}{dx^6} + y = \sin \tfrac{3}{2}x \sin \tfrac{1}{2}x$, $\qquad y = \dfrac{\cos 2x}{126} + \dfrac{x \sin x}{12} +$ C.F.

13. $\dfrac{d^4y}{dx^4} + 32\dfrac{dy}{dx} + 48y = xe^{-2x}$, $\quad y = \dfrac{e^{-2x}}{144}(x^3 + x^2) +$ C.F.

14. $\dfrac{d^4y}{dx^4} + 2\dfrac{d^2y}{dx^2} + y = x^2 \cos x$,

$\qquad y = \dfrac{x^3 \sin x}{12} + \dfrac{9x^2 - x^4}{48} \cos x +$ C.F.

15. $\dfrac{d^3y}{dx^3} - 2\dfrac{dy}{dx} + 4y = e^x \cos x$,

$\qquad y = \dfrac{xe^x}{20}(3 \sin x - \cos x) +$ C.F. (Compare Ex. IX., 18.)

16. $\left(\dfrac{d}{dx} + 1\right)^3 y = x^2 + x^{-1}$,

$\qquad y = e^{-x}(c_1 + c_2 x + c_3 x^2) + x^2 - 6x + 12 + e^{-x}\iiint \dfrac{e^x}{x} dx^3$.

17. $(D + b)^n y = \cos ax$,

$\qquad y = e^{-bx}(c_1 + c_2 x + \ldots + c_n x^{n-1})$
$\qquad\qquad + (a^2 + b^2)^{\frac{n}{2}} \cos\left(ax - n \cot^{-1}\dfrac{b}{a}\right)$.

18. Expand the integral $\int x^n e^x dx$ by the symbolic method.

$\qquad \dfrac{1}{D} x^n e^x = e^x [x^n - nx^{n-1} + n(n-1)x^{n-2} - \ldots] + c$.

19. Prove the following extension of Leibnitz' theorem: —

$\qquad \phi(D) uv = u \cdot \phi(D) v + Du \cdot \phi'(D) v + \dfrac{D^2 u}{2!} \cdot \phi''(D) v + \ldots$,

and show that it includes the extended form of integration by parts, Int. Calc., Art. 74.

20. In the equation connecting the perpendicular upon a tangent with the radius of curvature,

$$\rho = p + \frac{d^2 p}{d\phi^2},$$

(Diff. Calc., Art. 349), p and ϕ may be regarded as polar coordinates of the foot of the perpendicular. Hence show that, if the radius of curvature be given in the form $\rho = f(\phi)$, the equation of the pedal is

$$r = b\cos(\theta + a) + \frac{1}{D^2 + 1} f(\theta),$$

and interpret the complementary function (W. M. Hicks, *Messenger of Mathematics*, vol. vi. p. 95).

21. The radius of curvature of the cycloid being $\rho = 4a\cos\phi$, find the equation of the pedal at the vertex. $r = 2a\theta \sin\theta$.

CHAPTER VI.

LINEAR EQUATIONS WITH VARIABLE COEFFICIENTS.

XI.

The Homogeneous Linear Equation.

122. The linear equation

$$A_0 x^n \frac{d^n y}{dx^n} + A_1 x^{n-1} \frac{d^{n-1} y}{dx^{n-1}} + \ldots + A_n y = X,$$

in which the coefficient of each derivative is the product of a constant and a power of x whose exponent is the index of the derivative, is called the *homogeneous linear equation*. The operation expressed by each term of the first member is such that, when performed upon x^m, the result is a multiple of x^m; hence, if we put $y = x^m$ in the first member, the whole result will be the product of x^m and a constant factor involving m. Supposing then, in the first place, that the second member is zero, the equation will be satisfied if the value of m be so taken as to make the last-mentioned factor vanish. For example, if, in the equation

$$x^2 \frac{d^2 y}{dx^2} + 2x \frac{dy}{dx} - 2y = 0, \quad \ldots \quad (1)$$

we put $y = x^m$, the result is

$$[m(m-1) + 2m - 2]x^m = 0;$$

§ XI.] *THE HOMOGENEOUS LINEAR EQUATION.* 125

hence, if m satisfies the equation

$$m^2 + m - 2 = 0, \quad \ldots \ldots \quad (2)$$

x^m is an integral of the given equation. The roots of equation (2) are 1 and -2, giving two distinct integrals; hence, by Art. 93,

$$y = c_1 x + c_2 x^{-2}$$

is the complete integral of equation (1).

The Operative Symbol ϑ.

123. The homogeneous linear equation can be reduced to the form having constant coefficients by the transformation $x = e^\theta$. For, if $x = e^\theta$, we have (Diff. Calc., Art. 417)

$$x\frac{d}{dx} = \frac{d}{d\theta}, \qquad x^2\frac{d^2}{dx^2} = \frac{d}{d\theta}\left(\frac{d}{d\theta} - 1\right);$$

and, in general,

$$x^r\frac{d^r}{dx^r} = \frac{d}{d\theta}\left(\frac{d}{d\theta} - 1\right)\cdots\left(\frac{d}{d\theta} - r + 1\right);$$

so that in the transformation each term of the first member of the given equation gives rise to terms involving derivatives with respect to θ with constant coefficients only. Denoting $\dfrac{d}{d\theta}$ by D, the equation is thus reduced to the form

$$f(D)y = 0 \quad \ldots \ldots \quad (1)$$

in which f is an algebraic function having constant coefficients.

Now, if we put ϑ for the operative symbol $x\dfrac{d}{dx}$, the transforming equations become

$$x\frac{d}{dx} = \vartheta, \qquad x^2\frac{d^2}{dx^2} = \vartheta(\vartheta - 1),$$

and, in general,

$$x^r\frac{d^r}{dx^r} = \vartheta(\vartheta - 1)(\vartheta - 2)\ldots(\vartheta - r + 1);$$

and the result of transformation is

$$f(\vartheta)y = 0,^* \quad\ldots\ldots\ldots\ldots (2)$$

in which f denotes the same function as in equation (1), but x is still regarded as the independent variable. As an example of the transformation of an equation to the form (2), equation (1) of Art. 122 becomes

$$[\vartheta(\vartheta - 1) + 2\vartheta - 2]y = 0,$$

or

$$(\vartheta^2 + \vartheta - 2)y = 0.$$

124. The operator ϑ has the same relation to the function x^m that D has to e^{mx}; for we have

$$\vartheta x^m = mx^m, \quad \vartheta^2 x^m = m^2 x^m, \quad \ldots \quad \vartheta^r x^m = m^r x^m;$$

whence

$$f(\vartheta)x^m = f(m)x^m. \quad\ldots\ldots\ldots (1)$$

---

* The factors x and $\dfrac{d}{dx}$ of the symbol $x\dfrac{d}{dx}$ are non-commutative with one another, and the entire symbol, or ϑ, is non-commutative both with x and with D; but it is commutative with constant factors, and therefore is combined with them in accordance with the ordinary algebraic laws.

Thus the result of putting $y = x^m$ in the homogeneous linear equation
$$f(\vartheta)y = 0 \qquad (2)$$
is $f(m)x^m = 0$; whence
$$f(m) = 0. \qquad (3)$$

Accordingly, it will be noticed that the process of finding the function of m, as illustrated in Art. 122, is precisely the same as that of finding the function of ϑ, as illustrated in Art. 123.

If, now, the equation $f(m) = 0$ has n distinct roots, $m_1, m_2 \ldots m_n$, the complete integral of $f(\vartheta)y = 0$ is

$$y = C_1 x^{m_1} + C_2 x^{m_2} + \ldots + C_n x^{m_n}; \qquad (4)$$

the result being the same as that of substituting x for e^x in equation (3), Art. 95.

Cases of Equal and Imaginary Roots.

125. The modifications of the form of the integral, when $f(\vartheta) = 0$ has equal roots, or a pair of imaginary roots, may be derived from the corresponding changes in the case of the equation with constant coefficients. Thus, when $f(\vartheta) = 0$ has a double root equal to m, we find, by putting x in place of e^x, and consequently $\log x$ in place of x, in the results given in Art. 97, that the corresponding terms of the integral are

$$x^m(A + B \log x).$$

In like manner, when a triple root equal to m occurs, the corresponding terms are

$$x^m[A + B \log x + C(\log x)^2],$$

and so on.

Again, when $f(\vartheta) = 0$ has a pair of imaginary roots, $\alpha \pm \beta i$, we infer, from Art. 99, that the corresponding terms of the integral may be written

$$x^{\alpha}[A \cos(\beta \log x) + B \sin(\beta \log x)].$$

The Particular Integral.

126. The homogeneous linear equation, in which the second member is not zero, may be reduced to the form

$$f(\vartheta)y = X.$$

The complementary function, which is the integral of $f(\vartheta)y = 0$, is found by the method explained in the preceding articles.

The determination of the particular integral, which is symbolically expressed by $\dfrac{1}{f(\vartheta)} X$, may, by the resolution of $\dfrac{1}{f(\vartheta)}$ into partial fractions, be reduced to the evaluation of expression of the form

$$\frac{1}{\vartheta - a} X, \quad \frac{1}{(\vartheta - a)^2} X, \quad \text{etc.}$$

Compare Art. 105. The first of these expressions is the value of y in the equation

$$(\vartheta - a)y = X,$$

or

$$x \frac{dy}{dx} - ay = X,$$

a linear equation of the first order, whose integral is

$$x^{-a}y = \int x^{-a-1} X dx;$$

hence
$$\frac{1}{\vartheta - a} X = x^a \int x^{-a-1} X dx. \quad \ldots \ldots (1)$$

Again, applying the operation $\frac{1}{\vartheta - a}$ to both members of this equation, and reducing by means of the same equation,

$$\frac{1}{(\vartheta - a)^2} X = \frac{1}{\vartheta - a} x^a \int x^{-a-1} X dx = x^a \int x^{-1} \int x^{-a-1} X dx^2; \quad (2)$$

and, in general,

$$\frac{1}{(\vartheta - a)^r} X = x^a \int x^{-1} \int x^{-1} \int \ldots \int x^{-a-1} X dx^r. \quad \ldots (3)$$

127. Methods of operating with inverse symbols involving ϑ applicable to certain forms of the operand X, and analogous to those given in the preceding section for symbols involving D, might be deduced. The case of most frequent occurrence is that in which X is of the form x^a. From equation (1), Art. 124, it follows that, except when $f(a) = 0$,

$$\frac{1}{f(\vartheta)} x^a = \frac{1}{f(a)} x^a.$$

In the exceptional case, a is a root of $f(\vartheta) = 0$, and $f(\vartheta)$ is of the form $(\vartheta - a)^r \phi(\vartheta)$ where $\phi(a)$ does not vanish; hence

$$\frac{1}{(\vartheta - a)^r} \frac{1}{\phi(\vartheta)} x^a = \frac{1}{\phi(a)} \frac{1}{(\vartheta - a)^r} x^a.$$

But, by formula (3) of the preceding article,

$$\frac{1}{(\vartheta - a)^r} x^a = \frac{x^a}{r!} (\log x)^r.$$

130 *LINEAR EQUATIONS: VARIABLE COEFFICIENTS.* [Art. 127.

As an example, let the given equation be

$$x^2\frac{d^2y}{dx^2} + 2x\frac{dy}{dx} - 2y = x^3 + x;$$

the complementary function was found in Art. 122; for the particular integral, we have

$$y = \frac{1}{\vartheta^2 + \vartheta - 2}x^3 + \frac{1}{\vartheta - 1}\frac{1}{\vartheta + 2}x$$

$$= \frac{x^3}{10} + \frac{1}{3}\frac{1}{\vartheta - 1}x = \frac{x^3}{10} + \frac{x}{3}\int\frac{dx}{x}.$$

Hence

$$y = c_1x + c_2x^{-2} + \tfrac{1}{10}x^3 + \tfrac{1}{3}x\log x.$$

Symbolic Solutions.

128. The first member of any linear equation may be written symbolically

$$f(D, x)y.$$

In the case of the linear equation with constant coefficients, the operator is a function of D only. In the case of the homogeneous equation considered in the preceding articles, the operator is capable of expression as a function of the product xD which we denoted by ϑ. Examples occasionally occur in which $f(D, x)$ admits of expression as a function of some other single symbol which, like ϑ, involves D only in the first degree. In these cases, the equation is readily solved in the manner illustrated below.

Given the equation

$$\frac{d^2y}{dx^2} - 2bx\frac{dy}{dx} + b^2x^2y = 0.$$

Since $(D - bx)(Dy - bxy) = D^2y - bxDy - by - bxDy + b^2x^2y$, the equation may be written in the form

$$(D - bx)^2y + by = 0;$$

or, putting ζ for $D - bx$,

$$(\zeta^2 + b)y = 0,$$

in which the operator is expressed as a function of ζ. Resolving it into symbolic factors, we have

$$(\zeta - i\sqrt{b})(\zeta + i\sqrt{b})y = 0;$$

and the two terms of the integral satisfy respectively the equations

$$(\zeta - i\sqrt{b})y = 0 \quad \text{and} \quad (\zeta + i\sqrt{b})y = 0.$$

The first of these equations gives

$$(D - bx - i\sqrt{b})y_1 = 0,$$

or

$$\frac{dy_1}{y_1} = (bx + i\sqrt{b})dx;$$

and, integrating,

$$\log y_1 = \tfrac{1}{2}bx^2 + i\sqrt{b}x + c_1,$$

or

$$y_1 = C_1 e^{\tfrac{1}{2}bx^2}(\cos x\sqrt{b} + i \sin x\sqrt{b}).$$

In like manner, the second equation gives

$$y_2 = C_2 e^{\tfrac{1}{2}bx^2}(\cos x\sqrt{b} - i \sin x\sqrt{b}).$$

Adding, and changing the constants, as in Art. 99, we have, for the complete integral,

$$y = e^{\tfrac{1}{2}bx^2}(A \cos x\sqrt{b} + B \sin x\sqrt{b}).$$

When the second member of the given equation is a function of x, the particular integral is found by resolving the inverse symbol into partial fractions, as in Art. 126, the evaluation of each term depending on the solution of a linear equation of the first order.

129. The symbolic operator can sometimes be resolved into factors which are of the first degree with respect to D, but are not expressible in terms of any single operating symbol. In these cases, the factors are non-commutative; the equation can still be solved, but this circumstance materially alters the mode of solution. For example, the equation

$$\frac{d^2y}{dx^2} - (x^2 + x)\frac{dy}{dx} + (x^3 - 2x)y = X \quad . \quad . \quad . \quad (1)$$

may be written in the form

$$(D - x)(D - x^2)y = X; \quad . \quad . \quad . \quad . \quad (2)$$

for, by differentiation,

$$\left(\frac{d}{dx} - x\right)\left(\frac{dy}{dx} - x^2 y\right) = \frac{d^2y}{dx^2} - x^2 \frac{dy}{dx} - 2xy - x\frac{dy}{dx} + x^3 y.$$

The complementary function satisfies the equation

$$(D - x)(D - x^2)y = 0; \quad . \quad . \quad . \quad . \quad (3)$$

and it is evident that the solution of

$$(D - x^2)y = 0,$$

which is $y = Ce^{\frac{1}{3}x^3}$, satisfies equation (3). But, since we cannot reverse the order of the symbolic factors, equation (3) is *not* satisfied by the solution of $(D - x)y = 0$.

§ XI.]   NON-COMMUTATIVE SYMBOLIC FACTORS.   133

130. To solve equation (2), put

$$(D - x^2)y = v; \quad \ldots \ldots \quad (4)$$

then the equation becomes

$$(D - x)v = X, \quad \ldots \ldots \quad (5)$$

a linear equation of the first order for v. Solving equation (5), we have

$$v = e^{\frac{1}{2}x^2}\int e^{-\frac{1}{2}x^2}X dx + c_1 e^{\frac{1}{2}x^2};$$

and, substituting in equation (4), we have, by integration,

$$y = e^{\frac{1}{3}x^3}\int e^{-\frac{1}{3}x^3 + \frac{1}{2}x^2}\int e^{-\frac{1}{2}x^2}X dx^2 + c_1 e^{\frac{1}{3}x^3}\int e^{-\frac{1}{3}x^3 + \frac{1}{2}x^2}dx + c_2 e^{\frac{1}{3}x^3}.. \quad (6)$$

131. The solution of the general linear equation of the first order

$$(D + P)y = X$$

may (see Art. 34) be written in the symbolic form

$$y = \frac{1}{D + P}X = e^{-\int P dx}\int e^{\int P dx}X dx,$$

which includes the complementary function since the integral sign implies an arbitrary constant. In accordance with the same notation, the value of y, in equation (2), would be written

$$y = \frac{1}{D - x^2}\frac{1}{D - x}X,$$

which is at once reduced to the expression (6) by the above formula. It will be noticed that the factors must, in the inverse symbol, be written in the order inverse to that in which they occur in the direct symbol.

In obtaining this solution, the non-commutative character of the factors precluded us from a process analogous to the method of partial fractions, Art. 105; we have, in fact, only a solution analogous to equation (1) of Art. 104.

EXAMPLES XI.

Solve the following differential equations:—

1. $x^3\dfrac{d^3y}{dx^3} + 4x^2\dfrac{d^2y}{dx^2} - 2y = 0$, $\quad y = c_1 x^{-1} + c_2 x^{\sqrt{2}} + c_3 x^{-\sqrt{2}}$.

2. $x^3\dfrac{d^3y}{dx^3} + x\dfrac{d^2y}{dx^2} - 4\dfrac{dy}{dx} = 0$, $\quad y = c_1 x^3 + \dfrac{c_2}{x} + c_3$.

3. $2x^2\dfrac{d^2y}{dx^2} + 3x\dfrac{dy}{dx} - 3y = X$,

$$y = c_1 x + c_2 x^{-\frac{3}{2}} + \tfrac{2}{5}x\int\dfrac{X dx}{x^2} - \tfrac{2}{5}x^{-\frac{3}{2}}\int x^{\frac{1}{2}}X dx.$$

4. $x^2\dfrac{d^3y}{dx^3} - 2\dfrac{dy}{dx} = 0$, $\quad y = c_1 x^3 + c_2 + c_3 \log x$.

5. $x^2\dfrac{d^2y}{dx^2} + 4x\dfrac{dy}{dx} + 2y = e^x$, $\quad y = c_1 x^{-1} + c_2 x^{-2} + x^{-2}e^x$.

6. $x^2\dfrac{d^3y}{dx^3} + 3x\dfrac{d^2y}{dx^2} + 2\dfrac{dy}{dx} = x$,

$$y = A\cos\log x + B\sin\log x + C + \tfrac{1}{10}x^2.$$

7. $x^4\dfrac{d^3y}{dx^3} + 2x^3\dfrac{d^2y}{dx^2} - x^2\dfrac{dy}{dx} + xy = 1$,

$$y = x(A + B\log x) + Cx^{-1} + \tfrac{1}{4}x^{-1}\log x.$$

8. $x^4\dfrac{d^4y}{dx^4} + 6x^3\dfrac{d^3y}{dx^3} + 9x^2\dfrac{d^2y}{dx^2} + 3x\dfrac{dy}{dx} + y = 4x$,

$$y = (c_1 + c_2 \log x)\cos\log x + (c_3 + c_4 \log x)\sin\log x + x.$$

9. $x^3\dfrac{d^3y}{dx^3} + 2x^2\dfrac{d^2y}{dx^2} + 2y = 10\left(x + \dfrac{1}{x}\right)$,

$y = x(A\cos\log x + B\sin\log x + 5) + x^{-1}(C + 2\log x)$.

10. $x^3\dfrac{d^3y}{dx^3} + 2x^2\dfrac{d^2y}{dx^2} - x\dfrac{dy}{dx} + y = x\log x$,

$y = \dfrac{c_1}{x} + x\left[c_2 + c_3\log x - \dfrac{(\log x)^2}{8} + \dfrac{(\log x)^3}{12}\right]$.

11. $\dfrac{d^2y}{dx^2} + 4x\dfrac{dy}{dx} + 4x^2 y = 0$, $\quad y = c_1 e^{-x^2 - x\sqrt{2}} + c_2 e^{-x^2 + x\sqrt{2}}$.

12. Prove that

$$\vartheta^n x^r \log x = r^n x^r \log x + n r^{n-1} x^r.$$

13. Prove that

$$f(\vartheta) x^a V = x^a f(\vartheta + a) V$$

both when $f(\vartheta)$ is a direct, and when it is an inverse, symbol.

XII.

Exact Linear Equations.

132. Using accents to indicate the derivatives of y with respect to x, the linear equation of the nth order is

$$P_0 y^{(n)} + P_1 y^{(n-1)} + \ldots + P_{n-2} y'' + P_{n-1} y' + P_n y = X. \quad (1)$$

To ascertain the condition under which this represents an exact differential equation, and to find its integral when such is the case, we shall employ an extension of the method of Art. 84, which consists in successive subtractions of exact derivatives so chosen as to reduce the order of the remainder at each step until we arrive at a remainder which is, or is not, obviously exact. Since the second member of equation (1) is a function of x only, the equation is exact if, the subtractions being made from the first member, the coefficient of y vanishes in the remainder of the order zero, which contains no derivatives of y. When this condition is fulfilled, the sum of the expressions whose derivatives have been subtracted will be equal to $\int X dx + C$.

The first term of equation (1) shows that the first of these expressions is $P_0 y^{(n-1)}$, of which the derivative is

$$P_0 y^{(n)} + P_0' y^{(n-1)}.$$

Subtracting this from the first member, the remainder is

$$(P_1 - P_0') y^{(n-1)} + P_2 y^{(n-2)} + \ldots + P_n y;$$

or, putting

$$Q_1 = P_1 - P_0',$$

$$Q_1 y^{(n-1)} + P_2 y^{(n-2)} + \ldots + P_n y.$$

In like manner, the next expression whose derivative is to be subtracted is $Q_1 y^{(n-2)}$, the next remainder being

$$Q_2 y^{(n-2)} + P_3 y^{(n-3)} + \ldots + P_n y,$$

and so on, the values of Q_2, Q_3, etc., being

$$Q_2 = P_2 - Q_1', \quad Q_3 = P_3 - Q_2', \quad \text{etc.} \quad \ldots \quad (2)$$

The final remainder is $Q_n y$; and the condition of exactness is that this shall vanish, that is to say, $Q_n = 0$. If this condition be fulfilled, the integral will be

$$Q_0 y^{(n-1)} + Q_1 y^{(n-2)} + \ldots + Q_{n-2} y' + Q_{n-1} y = \int X dx + C \quad (3)$$

§ XII.] EXACT LINEAR EQUATIONS. 137

where $Q_0 = P_0$, $Q_1 = P_1 - P_0'$, $Q_2 = P_2 - P_1' + P_0''$, and in general,

$$Q_r = P_r - P_{r-1}' + P_{r-2}'' - \ldots \pm P_0^{(r)};$$

and the condition of direct integrability written at length is

$$Q_n = P_n - P_{n-1}' + P_{n-2}'' - \ldots \pm P_0^{(n)} = 0. \quad . \quad (4)$$

133. For example, to determine whether the equation

$$(x^3 - x)\frac{d^3y}{dx^3} + (8x^2 - 3)\frac{d^2y}{dx^2} + 14x\frac{dy}{dx} + 4y = \frac{2}{x^3}$$

is exact, we have, by the criterion, equation (4),

$$Q_3 = 4 - 14 + 16 - 6 = 0;$$

hence the equation is exact; and, forming the successive values of the coefficients Q by the equations (2), we find

$$(x^3 - x)\frac{d^2y}{dx^2} + (5x^2 - 2)\frac{dy}{dx} + 4xy = -\frac{1}{x^2} + c_1,$$

which is a first integral of the given equation.

Again, on applying the criterion to this result, we obtain $4x - 10x + 6x = 0$; hence it is also exact, and its integral is found, by the same process, to be

$$(x^3 - x)\frac{dy}{dx} + (2x^2 - 1)y = \frac{1}{x} + c_1x + c_2,$$

in which a second constant of integration is introduced.

This last result is not exact, for $2x^2 - 1 - (3x^2 - 1)$ is not equal to zero; but it is a linear equation of the first order, and its solution gives for the complete integral of the given equation,

$$xy\sqrt{(x^2-1)} = \sec^{-1}x + c_1\sqrt{(x^2-1)}$$
$$+ c_2 \log[x + \sqrt{(x^2-1)}] + c_3$$

134. The condition of direct integrability, equation (4), Art. 132, contains the rth derivative only of the coefficient of the rth derivative of y in equation (1); whence it is evident that the product

$$x^s \frac{d^r y}{dx^r}$$

is an exact derivative *when s is a positive integer less than r.* For example, $x^3 D^4 y$ is exact, because the fourth derivative of x^3 is zero; its integral is

$$x^3 D^3 y - 3x^2 D^2 y + 6x Dy - 6.$$

When s is negative, fractional, or an integer equal to or greater than r, a term of the form $x^s D^r y$, in equation (1), gives rise, in equation (4), to a term containing x^{s-r}. From this it is evident that, if, in the given equation, we group together the terms of the specified form in such a manner that $s - r$ has the same value for all the terms in a group, it is necessary, in order that the equation may be exact, that each group should separately constitute an exact derivative. If a single group be multiplied by x^m, and equation (4) be then formed, we shall have an equation by which m may be so determined that the group becomes exact; but, when the given equation consists of only one group, it becomes a homogeneous linear equation when multiplied by x^{r-s}, and it is more readily solved by the methods already given for such equations.

135. When an equation containing more than one such group of terms is not exact, it may happen that each group

§ XII.] *INTEGRATING FACTORS OF THE FORM* x^m.

becomes exact when multiplied by the same power of x. For example, the equation

$$2x^2(x+1)\frac{d^2y}{dx^2} + x(7x+3)\frac{dy}{dx} - 3y = X. \quad \ldots \quad (1)$$

contains two groups of terms, in one of which $s - r = 1$, and in the other $s - r = 0$. Multiplying by x^m, and then substituting in equation (4) of Art. 132, we have

$$-3x^m - 7(m+2)x^{m+1} - 3(m+1)x^m$$
$$+ 2(m+3)(m+2)x^{m+1} + 2(m+2)(m+1)x^m = 0,$$

which reduces to

$$(m+2)(2m-1)x^{m+1} + (m+2)(2m-1)x^m = 0, \quad (2)$$

the two terms in this equation respectively arising from the two groups in equation (1). If, now, the value of m can be so taken as to make each coefficient in equation (2) vanish, equation (1) becomes exact when multiplied by x^m. In this instance there are two such values of m; namely, -2 and $\frac{1}{2}$. Using the first value of m, we have the exact equation

$$2(x+1)\frac{d^2y}{dx^2} + \left(7 + \frac{3}{x}\right)\frac{dy}{dx} - \frac{3}{x^2}y = \frac{X}{x^2},$$

whose integral is

$$2(x+1)\frac{dy}{dx} + \left(5 + \frac{3}{x}\right)y = \int \frac{X}{x^2}dx + c_1; \quad \ldots \quad (3)$$

and, using the second value, we have the exact equation

$$2(x^{\frac{3}{2}} + x^{\frac{1}{2}})\frac{d^2y}{dx^2} + (7x^{\frac{1}{2}} + 3x^{-\frac{1}{2}})\frac{dy}{dx} - 3x^{-\frac{1}{2}}y = x^{\frac{1}{2}}X,$$

whose integral is

$$2x^{\frac{1}{2}}(x+1)\frac{dy}{dx} - 2x^{\frac{1}{2}}y = \int x^{\frac{1}{2}}Xdx - c_2. \quad \ldots \quad (4)$$

Having thus two first integrals of equation (1), its complete integral is found, by elimination of y' from equations (3) and (4), to be

$$5(x+1)y = c_1 x + c_2 x^{-\frac{3}{2}} + x\int\frac{X}{x^2}dx - x^{-\frac{3}{2}}\int x^{\frac{5}{2}}X dx. \quad (5)$$

Symbolical Treatment of Exact Linear Equations.

136. The result of a direct integration is, when regarded symbolically, equivalent to the resolution of the symbolic operator into factors, of which that most remote from the operand y is the simple factor D. For example, the two successive direct integrations effected in Art. 133 show that

$$(x^3 - x)D^3 + (8x^2 - 3)D^2 + 14xD + 4$$
$$= D^2[(x^3 - x)D + 2x^2 - 1];$$

and, from Art. 135, we infer the two results,

$$2x^2(x+1)D^2 + x(7x+3)D - 3$$
$$= x^2 D[2(x+1)D + 5 + 3x^{-1}]$$
$$= x^{-\frac{1}{2}}D[2x^{\frac{5}{2}}(x+1)D - 2x^{\frac{3}{2}}].$$

137. If, in a group of terms of the kind considered in Art. 134, m be the least value of r, and $q - m$ be the constant value of $s - r$, the group may be written

$$x^q(A_0 + A_1 xD + A_2 x^2 D^2 + \ldots)D^m y, \quad \ldots \quad (1)$$

where $A_0, A_1, \ldots$, are constant coefficients, and q may be negative or fractional. Using ϑ, as in Art. 123, to denote the operator xD, the expression in parenthesis may be reduced to the form $f(\vartheta)$, and the group to the form

$$x^q f(\vartheta) D^m y. \quad \ldots \ldots \ldots \quad (2)$$

It is shown in Art. 134 that, if m is not zero, and q is zero or a positive integer less than m, every term in the expression (1), and hence the whole expression (2), is an exact derivative. The symbolic transformation expressing the result, in this case, may be effected by means of the formula deduced below.

138. We have, by differentiation,

$$\frac{d}{dx} x \frac{dy}{dx} = x\frac{d^2y}{dx^2} + \frac{dy}{dx},$$

or

$$D\vartheta y = \vartheta D y + D y;$$

whence symbolically

$$\vartheta D = D(\vartheta - 1). \quad \ldots \ldots \ldots \quad (1)$$

Operating successively with ϑ upon both members, we derive

$$\vartheta^2 D = \vartheta D(\vartheta - 1) = D(\vartheta - 1)^2,$$
$$\vartheta^3 D = \vartheta D(\vartheta - 1)^2 = D(\vartheta - 1)^3;$$

and, in general,

$$\vartheta^r D = D(\vartheta - 1)^r.$$

Now, since $f(\vartheta)$ consists of terms of the form $A\vartheta^r$, it follows that

$$f(\vartheta) D = D f(\vartheta - 1).^* \quad \ldots \ldots \quad (2)$$

---

* The formula by which the homogeneous linear expression is reduced to the form $f(\vartheta)y$ is readily deduced from this formula. For equation (1) may be written

$$xD^2y = D(\vartheta - 1)y;$$

and, multiplying by x,

$$x^2 D^2 y = \vartheta(\vartheta - 1)y.$$

Changing the operand y to Dy, and using equation (2),

$$x^2 D^3 y = \vartheta(\vartheta - 1)Dy = D(\vartheta - 1)(\vartheta - 2)y.$$

Multiplying again by x,

$$x^3 D^3 y = \vartheta(\vartheta - 1)(\vartheta - 2)y;$$

and in like manner, we prove, in general,

$$x^r D^r y = \vartheta(\vartheta - 1)(\vartheta - 2) \ldots (\vartheta - r + 1)y.$$

Again, operating with each member of this equation upon D (which is equivalent to changing the operand from y to Dy),

$$f(\vartheta)D^2 = Df(\vartheta - 1)D = D^2f(\vartheta - 2).$$

In like manner,

$$f(\vartheta)D^3 = D^2f(\vartheta - 2)D = D^3f(\vartheta - 3);$$

and in general,

$$f(\vartheta)D^m = D^m f(\vartheta - m) \ldots \ldots \quad (3)$$

139. If q is a positive integer less than m, we can, by this formula, write

$$x^q f(\vartheta) D^m = x^q D^q f(\vartheta - q) D^{m-q};$$

whence

$$x^q f(\vartheta) D^m = \vartheta(\vartheta - 1) \ldots (\vartheta - q + 1) f(\vartheta - q) D^{m-q},$$

in which the expression for the group is reduced to the same form as when $q = 0$. We may now remove one or more of the factors of D^{m-q} to the extreme left of the symbol, thus effecting one or more, up to $m - q$, direct integrations, under the condition *that m is not zero, and that q has one of the values* $0, 1, 2 \ldots m - 1$.

The equation giving the result of $m - q$ integrations is

$$x^q f(\vartheta) D^m = D^{m-q}(\vartheta - m + q) \ldots (\vartheta - m + 1) f(\vartheta - m).$$

140. In every other case, the possibility of resolving the operator into factors of the required form depends upon the presence of a proper factor in $f(\vartheta)$. To show this, we have, by differentiation,

$$Dx^{q+1}y = x^{q+1}Dy + (q+1)x^q y;$$

whence, using Dx^{q+1} as a symbol of operation,

$$x^q(\vartheta + q + 1) = Dx^{q+1} \ldots \ldots \ldots \quad (1)$$

§ XII.]   CONDITIONS OF DIRECT INTEGRABILITY.   143

Now, if $-(q+1)$ is a root of the equation $f(\vartheta) = 0$, so that we can write
$$f(\vartheta) = (\vartheta + q + 1)\phi(\vartheta), \quad \ldots \quad (2)$$
we shall have
$$x^q f(\vartheta) D^m = D x^{q+1} \phi(\vartheta) D^m. \quad \ldots \quad (3)$$

We have thus a second condition* of direct integrability, and an expression for the result of integration.

141. If the first member of a differential equation be expressed in terms of the form $x^q f(\vartheta) D^m y$, the conditions given in Arts. 139 and 140 serve to show at once whether the equation can be made exact by multiplication by a power of x. For example, equation (1) of Art. 135, when written in the form considered, is

$$x^2(2\vartheta + 7)Dy + (2\vartheta + 3)(\vartheta - 1)y = X.$$

The first term becomes exact, in accordance with the first condition, when multiplied by x^{-2}; and the presence of the factor $(\vartheta - 1)$ shows that the second term is also made exact by the same factor. Hence, by equation (3), Art. 138, and equation (1), Art. 140, the symbolic operator may be written

$$x^2 D[(2\vartheta + 5) + x^{-1}(2\vartheta + 3)].$$

---

* This condition might be made to include that of the preceding article; for we might first, by means of equation (3), Art. 138, make the transformation
$$x^q f(\vartheta) D^m = x^{q-m} x^m D^m f(\vartheta - m),$$
and then the expression for $x^m D^m$, in terms of ϑ, which is
$$\vartheta(\vartheta - 1) \ldots (\vartheta - m + 1),$$
would, under the previous condition, contain the factor $\vartheta + q - m + 1$, which, in accordance with equation (1), should accompany x^{q-m}. But, since under no other condition would this happen, and since the factor would not appear in $f(\vartheta - m)$ unless $\vartheta + q + 1$ had been a factor of $f(\vartheta)$, this transformation is clearly unnecessary.

Again, both terms of the last factor fulfil the condition of Art. 140 when multiplied by $x^{\frac{1}{2}}$, and the expression becomes

$$2x^2 Dx^{-\frac{1}{2}} D(x^{\frac{3}{2}} + x^{\frac{1}{2}}).$$

The value of y obtained by performing upon X the inverse operations in the proper order is

$$y = \frac{1}{x^{\frac{3}{2}} + x^{\frac{1}{2}}} \int x^{\frac{1}{2}} \int \frac{Xdx}{2x^2} dx,$$

in which each integral sign implies an arbitrary constant. The expression is readily identified with that given in Art. 135.

It will be noticed that whenever an equation becomes exact when multiplied by either of two different powers of x, it is also susceptible of two successive direct integrations.

Examples XII.

Solve the following differential equations:—

1. $(x - 1)\dfrac{d^2y}{dx^2} + (x^2 + 1)\dfrac{dy}{dx} + 2xy = 0,$

$$e^{\frac{1}{2}x^2 + x}(x - 1)y = c_1 \int e^{\frac{1}{2}x^2 + x} dx + c_2.$$

2. $\dfrac{d^2y}{dx^2} + e^x\left(\dfrac{dy}{dx} + y\right) = e^x, \qquad e^x(y - 1) = c_1 \int e^{e^x} dx + c_2.$

3. $(1 - x^2)\dfrac{d^2y}{dx^2} - x\dfrac{dy}{dx} + y = 2x,$

$y = c_1 x + \sqrt{(1 - x^2)}(c_2 - \sin^{-1}x),$ or

$y = c_1 x + \sqrt{(x^2 - 1)}\{c_2 - \log[x + \sqrt{(x^2 - 1)}]\}.$

§ XII.] EXAMPLES.

4. $\dfrac{d^3y}{dx^3} + \cos x \dfrac{d^2y}{dx^2} - 2\sin x \dfrac{dy}{dx} - y\cos x = \sin 2x,$

$$y = e^{-\sin x}\int e^{\sin x}(c_1 x + c_2)dx + c_3 e^{-\sin x} - \dfrac{\sin x - 1}{2}.$$

5. $x\dfrac{d^3y}{dx^3} + (x^2 - 3)\dfrac{d^2y}{dx^2} + 4x\dfrac{dy}{dx} + 2y = 0,$

$$e^{\frac{x^2}{2}}\dfrac{y}{x^5} = c_1\int e^{\frac{x^2}{2}}\dfrac{dx}{x^5} + c_2\int e^{\frac{x^2}{2}}\dfrac{dx}{x^6} + c_3.$$

6. $x^2(x+2)\dfrac{d^2y}{dx^2} + x(x+3)\dfrac{dy}{dx} - 3y = X,$

$$\left(\dfrac{x}{x+2}\right)^{\frac{3}{2}} y = \int \dfrac{x^{\frac{3}{2}}}{(x+2)^{\frac{3}{2}}}\left(c_1 + \int \dfrac{X}{x^2}dx\right)dx + c_2.$$

7. $\sqrt{x}\dfrac{d^2y}{dx^2} + 2x\dfrac{dy}{dx} + 3y = x,$

$$e^{\frac{4}{3}x^{\frac{3}{2}}}\dfrac{y}{x} = \tfrac{1}{6}e^{\frac{4}{3}x^{\frac{3}{2}}} + c_1\int e^{\frac{4}{3}x^{\frac{3}{2}}}\dfrac{dx}{x^2} + c_2.$$

8. $(x^2 - x)\dfrac{d^2y}{dx^2} + 2(2x+1)\dfrac{dy}{dx} + 2y = 0,$

$$y = \dfrac{c_1}{(x-1)^5}(x^4 - 6x^2 + 2x - \tfrac{1}{3}) + \dfrac{x^3}{(x-1)^5}(c_2 - 4c_1 \log x).$$

9. $(x^2 - x)\dfrac{d^2y}{dx^2} - 2(x-1)\dfrac{dy}{dx} - 4y = 0,$

$$y = c_1(4x^3 - 2x^2 - \tfrac{2}{3}x - \tfrac{1}{3})$$
$$+ x^3(x-1)\left[c_2 - 4c_1 \log\dfrac{x}{x-1}\right].$$

10. Find three independent first integrals of the equation $y''' = X$.

$$y'' = \int X dx + c_1,\qquad xy'' - y' = \int xX dx + c_2,$$
$$x^2 y'' - 2xy' + 2y = \int x^2 X dx + c_3.$$

146 *LINEAR EQUATIONS: VARIABLE COEFFICIENTS.* [Art. 141.

11. Derive (α) the complete integral of $y''' = X$ from the above first integrals, and (β) the integral of $y^{iv} = X$ in like manner.

$$(\alpha), \quad 2y = x^2 \int X dx - 2x \int x X dx + \int x^2 X dx + \text{C. F.}$$

$$(\beta), \quad 6y = x^3 \int X dx - 3x^2 \int x X dx + 3x \int x^2 X dx - \int x^3 X dx + \text{C. F.}$$

12. Solve the equation

$$(2x + x^{\frac{1}{2}})\frac{d^2y}{dx^2} + 3\frac{dy}{dx} = X,$$

(α), as an equation of the first order for y'; (β), as an exact equation when multiplied by a proper power of x.

$$(\alpha), \quad y = A + \frac{Bx}{(2\sqrt{x}+1)^2} + \int (2\sqrt{x}+1)^{-3} \int \frac{(2\sqrt{x}+1)^2}{\sqrt{x}} X dx dx.$$

$$(\beta), \quad y = A + \frac{Bx}{(2\sqrt{x}+1)^2} + \frac{x}{(2\sqrt{x}+1)^2}\int \frac{2\sqrt{x}+1}{x^2} \int \sqrt{x} X dx dx.$$

13. Show that the equation

$$(2x^4 + 6x^{\frac{5}{2}})y^{iv} + (13x^3 + 41x^{\frac{3}{2}})y'''$$
$$+ (11x^2 + 54x^{\frac{1}{2}})y'' - (10x - 6x^{\frac{1}{2}})y' - 2y = X$$

may be written

$$(\vartheta + 1)^2(2\vartheta + 1)(\vartheta - 2)y$$
$$+ x^{\frac{1}{2}}(3\vartheta + 1)(2\vartheta + 3)(\vartheta + 2)Dy = X,$$

and find its integral.

$$y = \frac{1}{x}\int x^{\frac{2}{3}}(x^{\frac{1}{2}} + 3)^{\frac{2}{3}} dx \int \frac{dx}{x^{\frac{1}{2}}(x^{\frac{1}{2}}+3)^{\frac{5}{3}}} \int \frac{dx}{2x^{\frac{3}{2}}} \int X dx.$$

XIII.

The Linear Equation of the Second Order.

142. No general solution of the linear differential equation with variable coefficients exists when the order is higher than the first: there are, however, some considerations relating chiefly to equations of the second order which enable us to find the integral in particular cases, and to these we now proceed.

If a particular integral of the equation

$$\frac{d^2y}{dx^2} + P\frac{dy}{dx} + Qy = 0, \quad \ldots \quad (1)$$

in which P and Q are functions of x, be known, the complete integral, not only of this equation, but of the more general equation

$$\frac{d^2y}{dx^2} + P\frac{dy}{dx} + Qy = X, \quad \ldots \quad (2)$$

can be found. For let y_1 be the known integral of (1), and assume

$$y = y_1 v$$

in equation (2). Substituting, we have, for the determination of the new variable v,

$$\left. \begin{array}{c} y_1\dfrac{d^2v}{dx^2} + 2\dfrac{dy_1}{dx}\dfrac{dv}{dx} + \dfrac{d^2y_1}{dx^2}v \\[2pt] + Py_1\dfrac{dv}{dx} + P\dfrac{dy_1}{dx}v \\[2pt] + Qy_1v \end{array} \right\} = X. \quad \ldots \quad (3)$$

The coefficient of v in this equation vanishes by virtue of the hypothesis that y_1 satisfies equation (1); thus the equation becomes a linear equation of the first order for $\dfrac{dv}{dx}$ or v'. Hence v' may be determined; and then

$$y = y_1 \int v' dx + C_1 y_1$$

is the integral of equation (2), the other constant of integration being involved in the expression for v'.

143. As an illustration, let the given equation be

$$(1 - x^2)\dfrac{d^2y}{dx^2} + x\dfrac{dy}{dx} - y = x(1 - x^2)^{\frac{1}{2}},$$

in which, if the second member were zero, $y = x$ would obviously be a particular integral. Hence, assuming $y = xv$, and substituting,

$$x\dfrac{d^2v}{dx^2} + \left(2 + \dfrac{x^2}{1 - x^2}\right)\dfrac{dv}{dx} = x(1 - x^2)^{\frac{1}{2}},$$

or

$$\dfrac{dv'}{dx} + \left(\dfrac{2}{x} + \dfrac{x}{1 - x^2}\right)v' = (1 - x^2)^{\frac{1}{2}}.$$

Solving this equation, we have

$$\dfrac{x^2}{(1 - x^2)^{\frac{1}{2}}} v' = \int x^2 dx + c_1,$$

or

$$\dfrac{dv}{dx} = \tfrac{1}{3}x(1 - x^2)^{\frac{1}{2}} + c_1\dfrac{(1 - x^2)^{\frac{1}{2}}}{x^2};$$

and, integrating,

$$v = -\tfrac{1}{3}(1 - x^2)^{\frac{3}{2}} + c_1\left[\sin^{-1}x + \dfrac{(1 - x^2)^{\frac{1}{2}}}{x}\right] + c_2.$$

§ XIII.]   *A PARTICULAR INTEGRAL KNOWN.*   149

Hence
$$y = -\tfrac{1}{3}x(1-x^2)^{\frac{3}{2}} + c_1[x\sin^{-1}x + (1-x^2)^{\frac{1}{2}}] + c_2 x.$$

144. The simplification resulting from the substitution $y = y_1 v$ is due to the manner in which the constants enter the value of y in the complete integral. For we know that y is of the form
$$y = c_1 y_1 + c_2 y_2 + Y,$$

where y_1 and y_2 are independent particular integrals of the equation when the second member is zero, and Y is a particular integral when the second member is X. Hence the form of v is
$$v = c_1 + c_2 \frac{y_2}{y_1} + \frac{Y}{y_1},$$
and that of v' is
$$v' = c_2 \left(\frac{y_2}{y_1}\right)' + \left(\frac{Y}{y_1}\right)';$$

so that the equation determining v' must be a linear equation of the first order. In like manner, whatever be the degree of a linear equation, if a particular integral when the second member is zero be known, the order of the equation may be depressed by unity.

Expression for the Complete Integral in Terms of y_1.

145. The general equation for v', where y in the equation
$$\frac{d^2y}{dx^2} + P\frac{dy}{dx} + Qy = X \quad \ldots \ldots \quad (1)$$
is put equal to $y_1 v$, and y_1 satisfies
$$\frac{d^2y}{dx^2} + P\frac{dy}{dx} + Qy = 0, \quad \ldots \ldots \quad (2)$$

is [equation (3), Art. 142]

$$\frac{dv'}{dx} + \left(\frac{2}{y_1}\frac{dy_1}{dx} + P\right)v' = \frac{X}{y_1}.$$

Solving this linear equation of the first order, we have

$$y_1^2 e^{\int P dx} v' = \int y_1 e^{\int P dx} X dx + c_2;$$

and, since $y = y_1 v = y_1 \int v' dx$,

$$y = y_1 \int \frac{e^{-\int P dx}}{y_1^2} \int y_1 e^{\int P dx} X dx^2 + c_1 y_1 + c_2 y_1 \int \frac{e^{-\int P dx}}{y_1^2} dx \quad (3)$$

is the complete integral of equation (1) if y_1 is an integral of equation (2). Owing to the constants of integration implied in the integrals, the first term is, in reality, an expression for the complete integral: but the last two terms give a separate expression for the complementary function; that is to say, for the complete integral of equation (2).

146. Thus the complete integral of equation (2) may be written

$$y = c_1 y_1 + c_2 y_2$$

where

$$y_2 = y_1 \int \frac{e^{-\int P dx}}{y_1^2} dx. \quad \ldots \ldots \quad (4)$$

This expression may, in fact, represent any integral of equation (2); but, when the simplest values of the integrals involved in it are taken, it gives, when y_1 is known, the simplest independent integral; that is to say, the simplest integral which is not a mere multiple of y_1.

§ XIII.] RELATION BETWEEN THE TWO INTEGRALS. 151

For example, in the equation

$$x^2 \frac{d^2y}{dx^2} - 3x \frac{dy}{dx} + 4y = 0,$$

assuming, as in Art. 122, $y = x^m$, we have

$$m^2 - 4m + 4 = 0.$$

A case of equal roots arising, this gives but one integral of the simple form $y = x^m$, namely, $y_1 = x^2$. Now, in the given equation, $P = -\frac{3}{x}$; hence $e^{-\int P dx} = x^3$; and, substituting in equation (4), we have

$$y_2 = x^2 \int \frac{x^3}{x^4} dx = x^2 \log x$$

for the simplest independent integral.

147. The relation between the two independent integrals y_1 and y_2 may be put in a more symmetrical form. For equation (4), Art. 146, may be written

$$\frac{y_2}{y_1} = \int \frac{e^{-\int P dx}}{y_1^2} dx; \quad \ldots \ldots \quad (1)$$

whence, differentiating, we obtain

$$y_1 \frac{dy_2}{dx} - y_2 \frac{dy_1}{dx} = e^{-\int P dx}. \quad \ldots \ldots \quad (2)$$

This is a perfectly general relation between any two independent particular integrals of

$$\frac{d^2y}{dx^2} + P \frac{dy}{dx} + Qy = 0,$$

but it must be recollected that the value of the constant implied in the second member depends upon the form of the particular integrals y_1 and y_2. For this reason, the relation is better written

$$y_1 \frac{dy_2}{dx} - y_2 \frac{dy_1}{dx} = Ae^{-\int Pdx}. \quad \ldots \quad (3)$$

It will be noticed that, in this equation, the change of y_2 to my_2 multiplies A by m, but the change of y_2 to $y_2 + my_1$ does not affect A.

148. We may also, by introducing y_2, obtain a more symmetrical expression for the particular integral of the equation

$$\frac{d^2y}{dx^2} + P\frac{dy}{dx} + Qy = X$$

than that given in Art. 145. For, since by equation (1), Art. 147, $\frac{e^{-\int Pdx}}{y_1^2} dx = d\frac{y_2}{y_1}$, the particular integral in equation (3), Art. 145, may be written

$$Y = y_1 \int d\frac{y_2}{y_1} \int y_1 e^{\int Pdx} X dx,$$

which, by integration by parts, becomes

$$Y = y_2 \int y_1 e^{\int Pdx} X dx - y_1 \int y_2 e^{\int Pdx} X dx,$$

in which $\int Pdx$, in the exponential, is to be so taken as to satisfy equation (2); otherwise, the second member should be divided by the constant A defined by equation (3) of the preceding article.

Resolution of the Operator into Factors.

149. We have seen, in Art. 129, that, when the symbolic operator of a linear equation whose second member is zero is resolved into factors, the factor nearest the operand y gives, at once, an integral of the equation. Conversely, when an integral is known, the corresponding factor may be inferred; and, if the equation is of the second order, the other factor is found without difficulty.

For example, in the equation

$$(3-x)\frac{d^2y}{dx^2} - (9-4x)\frac{dy}{dx} + (6-3x)y = 0,$$

the fact that the sum of the coefficients is zero shows that e^x is an integral. The corresponding symbolic factor is $D - 1$, and accordingly the equation can be written

$$[(3-x)D - (6-3x)](D-1)y = 0.$$

The solution may now be completed as in Art. 130; thus, putting $v = (D-1)y$, we have

$$\frac{dv}{v} = \frac{3x-6}{x-3}dx,$$

the integral of which is

$$v = Ce^{3x}(x-3)^3.$$

Finally, solving the linear equation

$$(D-1)y = Ce^{3x}(x-3)^3,$$

we have the complete integral

$$y = Ae^x + Be^{3x}(4x^3 - 42x^2 + 150x - 183),$$

in which B is put for the constant $\tfrac{1}{8}C$.

150. In general, if y_1 denotes the known integral, and $D - \eta$ is the corresponding factor,

$$(D - \eta)y_1 = 0, \quad \text{or} \quad \frac{dy_1}{dx} - \eta y_1 = 0;$$

whence

$$\eta = \frac{1}{y_1} \frac{dy_1}{dx}. \quad \dots \dots \dots (1)$$

Now, in the case of the equation of the second order

$$(D^2 + PD + Q)y = 0, \quad \dots \dots \dots (2)$$

the other factor must be $D + P + \eta$ in order to make the first two terms of the expansion identical with those of equation (2); thus we have

$$(D + P + \eta)(D - \eta)y = 0, \quad \dots \dots \dots (3)$$

which, when expanded, is

$$D^2 y + PDy - \left(\frac{d\eta}{dx} + P\eta + \eta^2\right)y = 0. \quad \dots \dots (4)$$

The Related Equation of the First Order.

151. If, regarding η as an unknown function, we attempt to determine it by equating the coefficients of y in equations (2) and (4) of the preceding article, the result is

$$\frac{d\eta}{dx} + \eta^2 + P\eta + Q = 0. \quad \dots \dots \dots (1)$$

Hence, to any solution of this equation of the first order, there corresponds a solution of

$$\frac{d^2y}{dx^2} + P\frac{dy}{dx} + Qy = 0. \quad \dots \dots \dots (2)$$

§ XIII.] RELATED EQUATION OF THE FIRST ORDER. 155

Equation (1) is, in fact, merely the transformation of this equation when we put, as in the preceding article,

$$\eta = \frac{1}{y}\frac{dy}{dx}. \qquad \qquad (3)$$

Although of the first order, equation (1) is not so simple as equation (2), which has the advantage of being linear. In fact, the transformation just mentioned is advantageously employed in the solution of an equation of the form (1). See Art. 193. Since the complete integral of equation (2) is of the form

$$y = c_1 X_1 + c_2 X_2 \qquad \qquad (4)$$

where X_1 and X_2 are functions of x, that of equation (1) is of the form

$$\eta = \frac{c_1 X_1' + c_2 X_2'}{c_1 X_1 + c_2 X_2} = \frac{X_1' + c X_2'}{X_1 + c X_2}, \qquad \qquad (5)$$

which indicates the manner in which the arbitrary constant c enters the solution.

The particular integrals of (1) produced by giving different values to c correspond to independent integrals of equation (2), that is to say, integrals in which the ratio $c_2 : c_1$ has different values; the integrals in which $c = 0$ and $c = \infty$ in the expression (5) corresponding to the integrals X_1 and X_2 of equation (2).

The Transformation $y = vf(x)$.

152. If, in Art. 142, we replace y_1 by w_1, an arbitrary function of x, the result is that the equation

$$\frac{d^2 y}{dx^2} + P\frac{dy}{dx} + Qy = X \qquad \qquad (1)$$

is transformed, by the substitution

$$y = w_1 v, \qquad (2)$$

into

$$\frac{d^2v}{dx^2} + P_1\frac{dv}{dx} + Q_1 v = X_1, \qquad (3)$$

where

$$P_1 = \frac{2}{w_1}\frac{dw_1}{dx} + P, \qquad (4)$$

$$Q_1 = \frac{1}{w_1}\frac{d^2w_1}{dx^2} + \frac{P}{w_1}\frac{dw_1}{dx} + Q, \qquad (5)$$

$$X_1 = \frac{X}{w_1}. \qquad (6)$$

P_1, Q_1, and X_1 are here known functions of x; thus the equation remains linear when a transformation of the dependent variable of the form $y = vf(x)$ is made.

153. The arbitrary function w_1 can be so taken as to give to P_1 any desired value; thus, if P_1 is a given function of x, we have, from equation (4),

$$\frac{dw_1}{w_1} = \tfrac{1}{2}(P_1 - P)dx;$$

whence

$$w_1 = e^{\tfrac{1}{2}\int P_1 dx - \tfrac{1}{2}\int P dx}. \qquad (7)$$

Substituting in equations (5) and (6), we find, for the values of Q_1 and X_1, in terms of P_1,

$$Q_1 = Q + \tfrac{1}{4}(P_1^2 - P^2) + \tfrac{1}{2}\left(\frac{dP_1}{dx} - \frac{dP}{dx}\right) \qquad (8)$$

and

$$X_1 = X \frac{e^{\tfrac{1}{2}\int P dx}}{e^{\tfrac{1}{2}\int P_1 dx}}. \qquad (9)$$

§ XIII.]   THE TRANSFORMATION $y = vf(x)$.   157

These equations may be used in place of equations (5) and (6) when w_1 is given, P_1 being first found by means of equation (4).

154. Equation (4) may be written

$$P_1 = 2\frac{d}{dx}\log w_1 + P;$$

hence, when P is a rational algebraic fraction, if w_1 be taken of the form $e^{f(x)}$, where $f(x)$ is a rational algebraic function of x, P_1 will also be a rational fraction. From this and equation (8) it is manifest that, if the coefficients of the given differential equation are rational algebraic functions, those of the transformed equation will have the same character when w_1 is of the form $e^{f(x)}$, $f(x)$ being an algebraic function.

In particular, if the transformation is

$$y = e^{ax^m}v,$$

we have, since $\log w_1 = ax^m$,

$$P_1 = 2max^{m-1} + P;$$

and then, from equation (8),

$$Q_1 = m^2a^2x^{2m-2} + max^{m-1}P + m(m-1)ax^{m-2} + Q.$$

If, for example, this transformation be applied to the equation

$$\frac{d^2y}{dx^2} - 2bx\frac{dy}{dx} + b^2x^2y = 0,$$

we have $P = -2bx$ and $Q = b^2x^2$; whence

$$P_1 = 2max^{m-1} - 2bx,$$

$$Q_1 = m^2a^2x^{2m-2} - 2bmax^m + m(m-1)ax^{m-2} + b^2x^2.$$

If we put $m = 2$ and $a = \frac{1}{2}b$, P_1 vanishes, and Q_1 reduces to b; thus the transformed equation is

$$\frac{d^2v}{dx^2} + bv = 0,$$

of which the integral is

$$v = A \cos x\sqrt{b} + B \sin x\sqrt{b}.$$

Hence that of the given equation is

$$y = e^{\frac{1}{2}bx^2}(A \cos x\sqrt{b} + B \sin x\sqrt{b}),$$

agreeing with the solution otherwise found in Art. 128.

Removal of the Term containing the First Derivative.

155. If, in Art. 153, we take $P_1 = 0$, the transformed equation will not contain the first derivative. Distinguishing the corresponding values of w, Q, and X by the suffix zero, equation (7) gives

$$w_0 = e^{-\frac{1}{2}\int P dx}; \quad \dots \dots \dots (1)$$

so that the transformation is

$$y = v e^{-\frac{1}{2}\int P dx}, \quad \dots \dots \dots (2)$$

and the transformed equation is

$$\frac{d^2v}{dx^2} + Q_0 v = X_0, \quad \dots \dots \dots (3)$$

in which, by equations (8) and (9), Art. 153,

$$Q_0 = Q - \tfrac{1}{4}P^2 - \tfrac{1}{2}\frac{dP}{dx}. \quad \dots \dots (4)$$

$$X_0 = X e^{\frac{1}{2}\int P dx}. \quad \dots \dots \dots (5)$$

If the transformation $y = w_1 v$ is followed by the similar transformation $v = w_2 u$, where w_1 and w_2 are known functions of x, the effect is the same as that of the single transformation $y = w_1 w_2 u$, which is of the same form. It follows that the equations which are derivable from a given equation by transformations of the form $y = vf(x)$ constitute a system of equations transformable, in like manner, one into another. Among these equations there is a single equation of the form (3) which may thus be taken to represent the whole system. Accordingly equation (8), Art. 153, shows that the expression for Q_0, in equation (4), has an invariable value for all the equations of the system. The expression is therefore said to be an *invariant* for the transformation $y = vf(x)$.

156. One of the advantages of reducing an equation to the form (3), which may be called the *normal* form, is that, if any one of the equations of the system belongs to either of the classes for which we have general solutions, the equation in the normal form belongs to that class. For, in the first place, if, in any equation of the system, P and Q have constant values, equation (4) of the preceding article shows that Q_0 will also be constant. In the second place, if any one of the equations of the system is of the homogeneous form

$$\frac{d^2 y}{dx^2} + \frac{A}{x}\frac{dy}{dx} + \frac{B}{x^2}y = X,$$

putting $P = \frac{A}{x}$, and $Q = \frac{B}{x^2}$ in equation (4), we obtain

$$Q_0 = \frac{4B - A^2 + 2A}{4x^2};$$

hence the transformed equation is of the homogeneous form.

157. As an example of reduction to the normal form, let us take the equation

$$\frac{d^2y}{dx^2} - 2\tan x \frac{dy}{dx} - (a^2 + 1)y = 0.$$

Here $P = -2\tan x$; therefore, by equations (1) and (4), Art. 155,

$$w_0 = e^{\int \tan x\, dx} = \sec x,$$

and

$$Q_0 = -(a^2 + 1) - \tan^2 x + \sec^2 x = -a^2.$$

Thus the transformed equation is

$$\frac{d^2v}{dx^2} - a^2v = 0.$$

The integral of this is

$$v = c_1 e^{ax} + c_2 e^{-ax};$$

hence that of the given equation is

$$y = \sec x (c_1 e^{ax} + c_2 e^{-ax}).$$

Change of the Independent Variable.

158. If the independent variable be changed from x to z, z being a known function of x, the formulæ of transformation are

$$\frac{dy}{dx} = \frac{dy}{dz}\frac{dz}{dx},$$

and

$$\frac{d^2y}{dx^2} = \frac{d^2y}{dz^2}\left(\frac{dz}{dx}\right)^2 + \frac{dy}{dz}\frac{d^2z}{dx^2}.$$

Making these substitutions, the equation

$$\frac{d^2y}{dx^2} + P\frac{dy}{dx} + Qy = X \quad \ldots \ldots \quad (1)$$

is transformed into

$$\left(\frac{dz}{dx}\right)^2 \frac{d^2y}{dz^2} + \left(\frac{d^2z}{dx^2} + P\frac{dz}{dx}\right)\frac{dy}{dz} + Qy = X, \ldots \quad (2)$$

which is still linear, the coefficients being expressible as functions of z.

159. If it be possible to reduce a given equation by this transformation to the form with constant coefficients, it is evident, from equation (2), that we must have $\left(\frac{dz}{dx}\right)^2$ equal to the product of Q by a constant. For example, given the equation

$$(1 - x^2)\frac{d^2y}{dx^2} - x\frac{dy}{dx} + m^2y = 0,$$

in which $Q = \frac{m^2}{1 - x^2}$; if transformation to the required form be possible, it will be the result of putting $\frac{dz}{dx} = \frac{1}{\sqrt{(1 - x^2)}}$; whence $z = \sin^{-1} x$. Making the transformation, we obtain

$$\frac{d^2y}{dz^2} + m^2y = 0,$$

which is of the desired form. Its integral is

$$y = A \cos mz + B \sin mz;$$

hence that of the given equation is

$$y = A \cos m \sin^{-1} x + B \sin m \sin^{-1} x.$$

In like manner, if it be possible to reduce the equation to the homogeneous linear form, we must have $\left(\dfrac{dz}{dx}\right)^2$ equal to the product of Qz^2 by a constant. But this transformation succeeds only in the cases in which that considered above also succeeds; for it gives to log z the same value which the preceding one gives to z; accordingly it is equivalent to the latter transformation followed by the transformation $z = \log \zeta$, which is that by which we pass from the form with constant coefficients to the homogeneous form (see Art. 123).

160. We may, if we choose, so take z as to remove the term containing the first derivative. Equating to zero the coefficient of this term in equation (2), Art. 158, we find, for the required value of z,

$$z = \int e^{-\int P dx} dx.$$

Using this relation to express x as a function of z, the transformed equation is

$$\frac{d^2y}{dz^2} + Q\left(\frac{dx}{dz}\right)^2 y = X\left(\frac{dx}{dz}\right)^2.$$

Examples XIII.

Solve the following differential equations:—

1. $x\dfrac{d^2y}{dx^2} + (3 - x)\dfrac{dy}{dx} - 3y = 0,$

$y = c_1 e^x + c_2(x^3 + 3x^2 + 6x + 6).$

2. $\dfrac{d^2y}{dx^2} - x^2\dfrac{dy}{dx} + x(y - 1) = 0,$

$y = c_1 x + c_2 x \displaystyle\int e^{\frac{x^3}{3}}\dfrac{dx}{x^2} + 1.$

§ XIII.] EXAMPLES.

3. $\dfrac{d^2y}{dx^2} - x\dfrac{dy}{dx} + (x-1)y = X$,

$$y = c_1 e^x + c_2 e^x \int e^{\frac{x^2}{2} - 2x} dx + e^x \int e^{\frac{x^2}{2} - 2x} \int e^{-\frac{x^2}{2} + x} X dx^2.$$

4. $\dfrac{d^2y}{dx^2} - ax\dfrac{dy}{dx} + a^2(x-1)y = 0$,

$$y = c_1 e^{ax} + c_2 e^{ax} \int e^{\frac{ax^2}{2} - 2ax} dx.$$

5. $(a + x^2)\dfrac{d^2y}{dx^2} - 2x\dfrac{dy}{dx} + 2y = 0$, $\quad y = c_1(x^2 - a) + c_2 x$.

6. $\dfrac{d^3y}{dx^3} - x\dfrac{d^2y}{dx^2} - \dfrac{dy}{dx} + xy = 0$,

$$y = c_1 e^x + c_2 e^{-x} + c_3 \left(e^x \int e^{\frac{1}{2}x^2 - x} dx - e^{-x} \int e^{\frac{1}{2}x^2 + x} dx \right).$$

7. $\dfrac{d^2y}{dx^2} + \dfrac{4}{x}\dfrac{dy}{dx} + \dfrac{2y}{x^2} - m^2 y = 0$,

$$y = c_1 x^{-2} e^{mx} + c_2 x^{-2} e^{-mx}.$$

8. $x^3 \dfrac{d^3y}{dx^3} + x^2 \dfrac{d^2y}{dx^2} - 2x\dfrac{dy}{dx} + 2y = 0$,

$$y = c_1 x^2 + c_2 x + c_3 \left(x^2 \int x^{-3} e^{-x} dx - x \int x^{-2} e^{-x} dx \right).$$

9. $(2x^3 - a)\dfrac{d^2y}{dx^2} - 6x^2 \dfrac{dy}{dx} + 6xy = 0$,

$$y = c_1(x^3 + a) + c_2 x.$$

10. $\dfrac{d^2y}{dx^2} + x\dfrac{dy}{dx} + \dfrac{(x^2+1)^2}{4x^2} y = 0$, $\quad y = x^{\frac{1}{2}} e^{-\frac{x^2}{4}} (c_1 + c_2 \log x)$.

11. $x^2 \dfrac{d^2y}{dx^2} + (x - 4x^2)\dfrac{dy}{dx} + (1 - 2x + 4x^2) y = 0$,

$$y = e^{2x}(c_1 \cos \log x + c_2 \sin \log x).$$

12. $x^2 \dfrac{d^2y}{dx^2} - 2nx \dfrac{dy}{dx} + (n^2 + n + a^2x^2)y = 0$,

$$y = x^n(c_1 \cos ax + c_2 \sin ax).$$

13. $\dfrac{d^2y}{dx^2} + \tan x \dfrac{dy}{dx} + y \cos^2 x = 0, \qquad y = c_1 \sin(\sin x + c_2).$

14. $(a^2 + x^2) \dfrac{d^2y}{dx^2} + x \dfrac{dy}{dx} - m^2 y = 0$,

$$y = c_1[x + \sqrt{(a^2 + x^2)}]^m + c_2[x - \sqrt{(a^2 + x^2)}]^m.$$

15. $x^2 \dfrac{d^2y}{dx^2} - 2(x^2 + x) \dfrac{dy}{dx} + (x^2 + 2x + 2)y = 0$,

$$y = e^x(c_1 x^2 + c_2 x).$$

16. $\dfrac{d^2y}{dx^2} + \dfrac{2}{x} \dfrac{dy}{dx} + n^2 y = 0, \qquad y = c_1 \dfrac{\sin nx}{x} + c_2 \dfrac{\cos nx}{x}.$

17. $(1 - x^2) \dfrac{d^2y}{dx^2} - 2x \dfrac{dy}{dx} + \dfrac{a^2 y}{1 - x^2} = 0$,

$$y = c_1 \cos \dfrac{a}{2} \log \dfrac{1+x}{1-x} + c_2 \sin \dfrac{a}{2} \log \dfrac{1+x}{1-x}.$$

18. $\dfrac{d^2y}{dx^2} + \dfrac{2}{x} \dfrac{dy}{dx} + \dfrac{a^2}{x^4} y = 0, \qquad y = c_1 \cos \dfrac{a}{x} + c_2 \sin \dfrac{a}{x}.$

19. $\dfrac{d^2y}{dx^2} + (\tan x - 1)^2 \dfrac{dy}{dx} - n(n-1)y \sec^4 x = 0$,

$$y = c_1 e^{(n-1)\tan x} + c_2 e^{-n \tan x}.$$

20. $(a^2 + x^2)^2 \dfrac{d^2y}{dx^2} + 2x(a^2 + x^2) \dfrac{dy}{dx} + a^2 y = 0$,

$$y = \dfrac{c_1 x}{\sqrt{(a^2 + x^2)}} + \dfrac{c_2}{\sqrt{(a^2 + x^2)}}.$$

21. Derive equation (3), Art. 147, in the form

$$y_1 \dfrac{dy}{dx} - y \dfrac{dy_1}{dx} = A e^{-\int P dx}, \quad \text{from} \quad \dfrac{d^2y}{dx^2} + P \dfrac{dy}{dx} + Qy = 0,$$

by eliminating Q and integrating the result.

§ XIII.] EXAMPLES. 165

22. Find the symbolic resolution of D^2 corresponding to the integral x of the equation $D^2 y = 0$.

$$D^2 = \left(D + \frac{1}{x}\right)\left(D - \frac{1}{x}\right).$$

23. Find the symbolic resolution of $D^2 - 1$ corresponding to the integrals $\cosh x$ and $\sinh x$ of the equation $(D^2 - 1)y = 0$.

$$D^2 - 1 = (D + \tanh x)(D - \tanh x)$$
$$= (D + \coth x)(D - \coth x).$$

24. Show that the ratio s of two independent integrals of

$$\frac{d^2 y}{dx^2} + P\frac{dy}{dx} + Qy = 0$$

satisfies the differential equation of the third order

$$\frac{s'''}{s'} - \frac{3}{2}\left(\frac{s''}{s'}\right)^2 = 2Q_0,$$

where Q_0 is the function defined in Art. 155.

25. Show that, if P be expressed in terms of z, the equation of Art. 160 may be written

$$\left(\int P dz\right)^2 \frac{d^2 y}{dz^2} + Qy = X.$$

26. Prove that, in the equation

$$\frac{d^2 y}{dx^2} + P\frac{dy}{dx} + Qy = 0,$$

the function

$$\left(2PQ + \frac{dQ}{dx}\right)Q^{-\frac{3}{2}}$$

is an invariant with respect to the transformation $z = \phi(x)$.

CHAPTER VII.

SOLUTIONS IN SERIES.

XIV.

Development of the Integral of a Differential Equation in Series.

161. In many cases, the only solution of a given differential equation obtainable is in the form of a development of the dependent variable y, in the form of an infinite series involving powers of the independent variable x. Moreover, such a development may be desired, even when the relation between x and y is otherwise expressible. If we assume the series to proceed by integral powers of x, an obvious method by which successive terms could generally be found is as follows. Supposing the equation to be of the nth order, and assuming, for the n arbitrary constants, the initial values corresponding to $x = 0$ of y and its derivatives, up to and inclusive of the $(n-1)$th, the differential equation serves to determine the value of $\frac{d^n y}{dx^n}$ when $x = 0$. Differentiating the given equation, we have an equation containing $\frac{d^{n+1} y}{dx^{n+1}}$, which, in like manner, serves to determine its value when $x = 0$, and so on. Thus, writing out the value of y in accordance with Maclaurin's theorem, we have the values of the successive coefficients in terms of n arbitrary constants.

162. It would usually be impossible to obtain, in the manner described above, the general term of the series. We shall therefore consider only the case of the linear equation (and such as can be reduced to a linear form), in which case we have a method, now to be explained, which allows us to assume the series in a more general form, and, at the same time, enables us to find the law of formation of the successive coefficients.

Since we know the form of the complete integral of a linear equation to be

$$y = c_1 y_1 + c_2 y_2 + \ldots + c_n y_n + Y,$$

our problem now is the more definite one of developing in series the independent integrals $y_1, y_2 \ldots y_n$, of the equation when the second member is zero, and the particular integral Y of the equation when the second member is a function of x. No arbitrary constants, it will be noticed, will now occur in the coefficients of the required series, except the single arbitrary constant factor in the case of each independent integral.

Development of the Independent Integrals of a Linear Equation whose Second Member is Zero.

163. We have seen, in Art. 122, that if, in the first member of a homogeneous linear equation whose second member is zero, we put $y = Ax^m$, the result is an expression containing a single power of x; so that, by putting the coefficient of this power equal to zero, we have an equation for determining m in such a manner that $y = Ax^m$ satisfies the differential equation, A being an arbitrary constant.

If we make the same substitution in any linear equation whose coefficients are rational algebraic functions of x, the result will contain several powers of x. Let us, for the present, suppose that it contains two powers of x, and also that the differential equation is of the second order. The term containing $\frac{d^2y}{dx^2}$ in the differential equation will produce at least one term, in the result of substitution, involving m in the second degree; hence at least one of the coefficients of the two powers of x will be of the second degree in m. Let $x^{m'}$ and $x^{m'+s}$, where s may have any value, positive or negative, be the two powers of x, and let the coefficient of $x^{m'}$ be of the second degree. Now let m be so determined that the coefficient of $x^{m'}$ shall vanish, and suppose the quadratic equation for this purpose to have real roots. Selecting either of the two values of m, the coefficient of $x^{m'+s}$ will, of course, not in general vanish.

Suppose, now, that we put for y, in the first member of the differential equation, the expression $A_0 x^m + A_1 x^{m+s}$, the result will contain, in addition to the previous result, a new binomial containing A_1, and involving the powers $x^{m'+s}$ and $x^{m'+2s}$; the entire coefficient of $x^{m'+s}$ will now contain A_0 and A_1, and may be made to vanish by properly determining the ratio of the assumed constants A_1 and A_0. In like manner, if we assume for y the infinite series

$$y = A_0 x^m + A_1 x^{m+s} + A_2 x^{m+2s} + \ldots,$$

or

$$y = \Sigma_0^\infty A_r x^{m+rs},$$

we can successively cause the coefficients in the result of substitution to vanish by properly determining the ratio of consecutive coefficients in the assumed series. If the series thus obtained is convergent, it defines an integral of the given equation; and, since in the case supposed there were two values of m determined, we have, in general, two integrals. If s be positive, the series will proceed by ascending powers, and, if s be negative, by descending powers, of x.

§ XIV.] DETERMINATION OF THE COEFFICIENTS.

164. For example, let the given equation be

$$\frac{d^2y}{dx^2} - x\frac{dy}{dx} - py = 0. \quad\quad\quad (1)$$

The result of putting $A_0 x^m$ for y in the first member is

$$m(m-1)A_0 x^{m-2} - (m+p)A_0 x^m. \quad\quad (2)$$

The first term, which is of the second degree with respect to m, will vanish if we put

$$m(m-1)A_0 = 0. \quad\quad\quad (3)$$

The exponent of x in this term, or m', is $m-2$, and the other exponent, or $m'+s$, is m; whence $s=2$. We therefore assume the ascending series

$$y = \Sigma_0^\infty A_r x^{m+2r},$$

and, substituting in equation (1), we have

$$\Sigma_0^\infty \{(m+2r)(m+2r-1)A_r x^{m+2r-2}$$
$$- (m+2r+p)A_r x^{m+2r}\} = 0, \quad (4)$$

in which r has all integral values from 0 to ∞.

In this equation, the coefficient of each power of x must vanish; hence, equating to zero, the coefficient of x^{m+2r-2}, we have

$$(m+2r)(m+2r-1)A_r - (m+2r-2+p)A_{r-1} = 0. \quad (5)$$

When $r=0$, this reduces to equation (3) and gives

$$m=0 \quad \text{or} \quad m=1;$$

and when $r>0$, it may be written

$$A_r = \frac{m+2r-2+p}{(m+2r)(m+2r-1)} A_{r-1}, \quad\quad (6)$$

which expresses the relation between any two consecutive coefficients.

When $m = 0$, this relation becomes

$$A_r = \frac{p + 2r - 2}{2r(2r - 1)} A_{r-1};$$

whence, giving to r the successive values 1, 2, 3 ..., we have

$$A_1 = \frac{p}{1 \cdot 2} A_0,$$

$$A_2 = \frac{p + 2}{3 \cdot 4} A_1 = \frac{p(p + 2)}{4!} A_0,$$

$$A_3 = \frac{p + 4}{5 \cdot 6} A_2 = \frac{p(p + 2)(p + 4)}{6!} A_0.$$

. . . .

The resulting value of y is

$$y = A_0 \left[1 + p\frac{x^2}{2!} + p(p + 2)\frac{x^4}{4!} + p(p + 2)(p + 4)\frac{x^6}{6!} + \ldots \right]. \quad (7)$$

Again, giving to m its other value 1, the relation (6) between consecutive coefficients becomes

$$B_r = \frac{p + 2r - 1}{(2r + 1)2r} B_{r-1};$$

whence

$$B_1 = \frac{p + 1}{2 \cdot 3} B_0,$$

$$B_2 = \frac{p + 3}{4 \cdot 5} B_1 = \frac{(p + 1)(p + 3)}{5!} B_0,$$

$$B_3 = \frac{p + 5}{6 \cdot 7} B_2 = \frac{(p + 1)(p + 3)(p + 5)}{7!} B_0;$$

and the resulting value of y is

$$y = B_0 \left[x + (p + 1)\frac{x^3}{3!} + (p + 1)(p + 3)\frac{x^5}{5!} + \ldots \right]. \quad (8)$$

§ XIV.]   CONVERGENCY OF THE SERIES.   171

Denoting the series in equations (7) and (8), both of which are converging for all values of x, by y_1 and y_2, the complete integral of equation (1) is

$$y = A_0 y_1 + B_0 y_2. \quad \ldots \ldots \ldots \quad (9)$$

165. It will be noticed that the rule which requires us to take, for the determination of m, that term of the expression (2) which is of the second degree in m was necessary to enable us to obtain two independent integrals. But there is a more important reason for the rule; for, if we disregard it, we obtain a divergent series. For example, in the present instance, if we employ the other term of expression (2), Art. 164, thus obtaining

$$m = -p \quad \text{and} \quad s = -2,$$

the resulting series is

$$y = A_0 x^{-p} \left[1 - \frac{p(p+1)}{2} x^{-2} + \frac{p(p+1)(p+2)(p+3)}{2 \cdot 4} x^{-4} - \ldots \right].$$

The ratio of the $(r+1)$th to the rth term is

$$-\frac{(p+2r-2)(p+2r-1)}{2r} x^{-2};$$

and this expression increases without limit as r increases, whatever be the value of x. Hence the series ultimately diverges for all values of x.

When both terms in the expression corresponding to (2) are of the second degree in m, we can obtain two series in descending powers of x as well as two in ascending powers; and, in such cases, the descending series will be convergent for values of x greater than unity, and the ascending series will be convergent for values less than unity.

The Particular Integral.

166. When the second member of a linear equation is a power of x, the method explained in the preceding articles serves to determine the complementary function, and the particular integral may be found by a similar process. Thus, if the equation is

$$\frac{d^2y}{dx^2} - x\frac{dy}{dx} - py = x^{\frac{1}{2}},$$

the complementary function is the value of y found in Art. 164. To obtain the particular integral, we assume for y the same form of series as before, and the result of substitution is the same as equation (4), Art. 164, except that the second member is $x^{\frac{1}{2}}$ instead of zero. Equation (5) thus remains unaltered, while, in place of equation (3), we have

$$m(m-1)A_0 x^{m-2} = x^{\frac{1}{2}}.$$

This equation requires us to put

whence
$$m - 2 = \tfrac{1}{2}, \quad \text{and} \quad m(m-1)A_0 = 1;$$
$$m = \tfrac{5}{2}, \quad \text{and} \quad A_0 = \tfrac{4}{15}.$$

The relation (6) between consecutive coefficients now becomes

$$A_r = \frac{p + 2r + \tfrac{1}{2}}{(2r + \tfrac{3}{2})(2r + \tfrac{5}{2})} A_{r-1},$$

or

$$A_r = \frac{2(2p + 4r + 1)}{(4r + 3)(4r + 5)} A_{r-1};$$

hence

$$A_1 = \frac{2(2p + 5)}{7 \cdot 9} A_0,$$

$$A_2 = \frac{2(2p + 9)}{11 \cdot 13} A_1 = \frac{2^2(2p + 5)(2p + 9)}{7 \cdot 9 \cdot 11 \cdot 13} A_0;$$

. . . .

and the particular integral is

$$Y = \frac{4}{15}x^{\frac{5}{2}}\left[1 + \frac{2(2p+5)}{7\cdot 9}x^2 + \frac{2^2(2p+5)(2p+9)}{7\cdot 9\cdot 11\cdot 13}x^4 + \ldots\right].$$

If the second member contained two or more terms, each of them would give rise to a series, and the sum of these series would constitute the particular integral.

Binomial and Polynomial Equations.

167. If we group together the terms of a linear equation whose coefficients are rational algebraic functions of x in the manner explained in Art. 134, we can, by multiplying by a power of x, and employing the notation $x\dfrac{d}{dx} = \vartheta$, put the equation in the form

$$f_1(\vartheta)y + x^{s_1}f_2(\vartheta)y + x^{s_2}f_3(\vartheta)y + \ldots = 0, \quad . \quad . \quad (1)$$

in which $s_1, s_2 \ldots$ are all positive, or, if we choose, all negative. The result of putting $A_0 x^m$ for y in the first member is

$$A_0 f_1(m)x^m + A_0 f_2(m)x^{m+s_1} + A_0 f_3(m)x^{m+s_2} + \ldots . \quad (2)$$

Equations may be classified as *binomial*, *trinomial*, etc., according to the number of terms they contain, when written in the form (1), or, what is the same thing, the number of terms in the result of substitution (2). Thus, the equation solved in Art. 164 is a binomial equation.

In the general case, the process of solving in series is similar to that employed in Art. 164, the form which it is necessary to assume for the series being

$$y = \Sigma_0^\infty A_r x^{m+rs},$$

where s is the greatest number, integral or fractional, which is contained a whole number of times in each of the quantities s_1, s_2, etc. As before, m is taken to be a root of the equation

$f_1(m) = 0$, and A_0 is arbitrary; but, when the coefficient of the general term in the complete result of substitution is equated to zero, the relation found between the assumed coefficients A_0, A_1, A_2, etc., involves three or more of them, so that each is expressed in terms of two or more of the preceding ones. We can thus determine as many successive coefficients as we please, but cannot usually express the general term of the series.

We shall, in what follows, confine our attention to binomial equations of the second order.

Finite Solutions.

168. It sometimes happens that the series obtained as the solution of a binomial equation terminates by reason of the occurrence of the factor zero in the numerator of one of the coefficients, so that we have a finite solution of the equation. For example, let the given equation be

$$\frac{d^2y}{dx^2} + a\frac{dy}{dx} - 2\frac{y}{x^2} = 0. \quad \ldots \ldots \quad (1)$$

This is obviously a binomial equation in which $s = 1$; hence, putting

$$y = \Sigma_0^\infty A_r x^{m+r},$$

we have

$$\Sigma_0^\infty \{[(m+r)(m+r-1) - 2]A_r x^{m+r-2} + a(m+r)A_r x^{m+r-1}\} = 0.$$

Equating to zero the coefficient of x^{m+r-2}, we have

$$(m+r+1)(m+r-2)A_r + a(m+r-1)A_{r-1} = 0,$$

which, when $r = 0$, gives

$$(m+1)(m-2)A_0 = 0; \quad \ldots \ldots \quad (2)$$

§ XIV.]   FINITE SOLUTIONS.   175

and, when $r > 0$,

$$A_r = -a\frac{m+r-1}{(m+r+1)(m+r-2)}A_{r-1}, \quad \ldots \quad (3)$$

The roots of equation (2) are $m = -1$ and $m = 2$; taking $m = -1$, the relation (3) becomes

$$A_r = -a\frac{r-2}{r(r-3)}A_{r-1}, \quad \ldots \ldots \quad (4)$$

in which, putting $r = 1$, and $r = 2$, we have

$$A_1 = -a\frac{-1}{1(-2)}A_0,$$

$$A_2 = -a\frac{0}{2(-1)}A_1 = 0.$$

All the following coefficients may now be taken equal to zero,*

---

* In general, when one of the coefficients vanishes, the subsequent coefficients in the assumed series $\Sigma_0^\infty A_r x^{m+rs}$ *must* vanish; in other words, the value of y can contain no other terms whose exponents are of the form $m + rs$. But, in the present case, the assumed form is $y = \Sigma_0^\infty A_r x^{r-1}$; and this includes the powers $x^2, x^3 \ldots$ which we know to be of possible occurrence since the other value of m in this case is 2. Accordingly, if we continue the series, it recommences with the term containing x^2. Thus, putting $r = 3$ in equation (4), we obtain

$$A_3 = -a\frac{1}{3.0}A_2 = \frac{0}{0},$$

which is indeterminate; then, putting $r = 4, 5$, etc., we have

$$A_4 = -a\frac{2}{4.1}A_3, \quad A_5 = -a\frac{3}{5.2}A_4 = a^2\frac{3}{4.5}A_3, \quad \text{etc.}$$

Thus, the assumed form $y = \Sigma_0^\infty A_r x^{r-1}$ really includes, in this case, the complete integral

$$y = A_0\left(\frac{1}{x} - \frac{a}{2}\right) + A_3 x^2\left(1 - \frac{2}{4}ax + \frac{3}{4.5}a^2x^2 - \frac{4}{4.5.6}a^3x^3 + \ldots\right).$$

so that we have the finite solution *

$$A_0 y_1 = A_0 x^{-1}\left(1 - \frac{a}{2}x\right) = A_0\left(\frac{1}{x} - \frac{a}{2}\right).$$

169. For the other solution, taking $m = 2$, the relation (3) becomes

$$B_r = -a\frac{r+1}{(r+3)r}B_{r-1};$$

whence

$$B_1 = -a\frac{2}{1.4}B_0,$$

$$B_2 = -a\frac{3}{2.5}B_1 = a^2\frac{3}{4.5}B_0,$$

$$B_3 = -a\frac{4}{3.6}B_2 = -a^3\frac{4}{4.5.6}B_0.$$

. . . .

Hence

$$B_0 y_2 = B_0 x^2\left(1 - \frac{2}{4}ax + \frac{3}{4.5}a^2x^2 - \frac{4}{4.5.6}a^3x^3 + \ldots\right);$$

and the complete integral is

$$y = A_0 \frac{2-ax}{2x} + B_0 x^2\left(1 - \frac{2}{4}ax + \frac{3}{4.5}a^2x^2 - \ldots\right).$$

170. Since we have, in this case, a finite integral of a linear equation of the second order, namely,

$$y_1 = \frac{2-ax}{2x},$$

---

* In like manner, if, in a trinomial equation, the coefficients between which the relation exists are consecutive, a finite solution will occur when two consecutive coefficients vanish.

§ XIV.] EXAMPLES. 177

equation (4), Art. 146, gives the independent integral

$$y_2' = \frac{2 - ax}{2x} \int_0^x \frac{4x^2 e^{-ax}}{(2 - ax)^2} dx.$$

We must therefore have $y_2' = Ay_1 + By_2$ where y_1 and y_2 are the integrals found in the preceding articles, and the constants A and B have particular values to be determined. Since both y_2' and y_2 vanish when $x = 0$, while y_1 does not, we shall have $A = 0$; and, comparing the lowest terms of the development of the integral with the series y_2, we find $B = \frac{1}{3}$; hence

$$\frac{2 - ax}{x} \int_0^x \frac{x^2 e^{-ax}}{(2 - ax)^2} dx = \frac{x^2}{6} \left[1 - \frac{2}{4} ax + \frac{3}{4 \cdot 5} a^2 x^2 - \cdots \right].$$

EXAMPLES XIV.

Integrate in series the following differential equations : —

1. $x \dfrac{d^2 y}{dx^2} + (x + n) \dfrac{dy}{dx} + (n + 1)y = 0,$

$$y = A \left(n - (n + 1)x + (n + 2)\frac{x^2}{2!} - (n + 3)\frac{x^3}{3!} + \cdots \right)$$

$$+ Bx^{1-n} \left(1 + \frac{2}{n - 2}x + \frac{3}{(n - 2)(n - 3)}x^2 \right.$$

$$\left. + \frac{4}{(n - 2)(n - 3)(n - 4)}x^3 + \cdots \right).$$

2. $\dfrac{d^2 y}{dx^2} + xy = 0,$

$$y = A \left(1 - \frac{1}{3!}x^3 + \frac{1 \cdot 4}{6!}x^6 - \frac{1 \cdot 4 \cdot 7}{9!}x^9 + \cdots \right)$$

$$+ B \left(x - \frac{2}{4!}x^4 + \frac{2 \cdot 5}{7!}x^7 - \cdots \right).$$

3. $2x^2\dfrac{d^2y}{dx^2} - x\dfrac{dy}{dx} + (1 - x^2)y = x^2,$

$y = Ax\left(1 + \dfrac{x^2}{2.5} + \dfrac{x^4}{2.4.5.9} + \dfrac{x^6}{2.4.6.5.9.13} + \cdots\right)$

$\qquad + Bx^{\frac{1}{2}}\left(1 + \dfrac{x^2}{2.3} + \dfrac{x^4}{2.4.3.7} + \dfrac{x^6}{2.4.6.3.7.11} + \cdots\right)$

$\qquad\qquad + \dfrac{x^2}{1.3} + \dfrac{x^4}{1.3.3.7} + \dfrac{x^6}{1.3.5.3.7.11} + \cdots$

4. $x\dfrac{d^2y}{dx^2} + 2\dfrac{dy}{dx} + a^3x^2y = 2,$

$y = A\left(1 - \dfrac{2}{4!}a^3x^3 + \dfrac{2.5}{7!}a^6x^6 - \cdots\right)$

$\qquad + Bx^{-1}\left(1 - \dfrac{1}{3!}a^3x^3 + \dfrac{1.4}{6!}a^6x^6 + \cdots\right)$

$\qquad + x\left(1 - 2\dfrac{3}{5!}a^3x^3 + 2\dfrac{3.6}{8!}a^6x^6 - 2\dfrac{3.6.9}{11!}a^9x^9 + \cdots\right).$

5. $\dfrac{d^2y}{dx^2} + ax^2y = 1 + x,$

$y = A\left(1 - \dfrac{ax^4}{3.4} + \dfrac{a^2x^8}{3.4.7.8} - \dfrac{a^3x^{12}}{3.4.7.8.11.12} + \cdots\right)$

$\qquad + Bx\left(1 - \dfrac{ax^4}{4.5} + \dfrac{a^2x^8}{4.5.8.9} - \dfrac{a^3x^{12}}{4.5.8.9.12.13} + \cdots\right)$

$+ \dfrac{x^2}{2}\left(1 - \dfrac{ax^4}{5.6} + \dfrac{a^2x^8}{5.6.9.10} - \cdots\right) + \dfrac{x^3}{6}\left(1 - \dfrac{ax^4}{6.7} + \dfrac{a^2x^8}{6.7.10.11} - \cdots\right).$

6. $x\dfrac{d^2y}{dx^2} + (x + n)\dfrac{dy}{dx} + (n - 1)y = x^{1-n},$

$y = A\left(1 - \dfrac{n-1}{n}\dfrac{x}{1!} + \dfrac{n-1}{n+1}\dfrac{x^2}{2!} - \dfrac{n-1}{n+2}\dfrac{x^3}{3!} + \cdots\right) + Bx^{1-n}$

$\qquad + \dfrac{x^{2-n}}{2-n}\left(1 - \dfrac{1}{3-n}\dfrac{x}{2} + \dfrac{1}{(3-n)(4-n)}\dfrac{x^2}{3} - \cdots\right).$

§ XIV.] EXAMPLES.

7. $x^4 \dfrac{d^2y}{dx^2} + x \dfrac{dy}{dx} + y = 0$,

$$y = A\left(1 - \frac{1}{3!}x^{-2} - \frac{1}{5!}x^{-4} - \frac{1\cdot3}{7!}x^{-6} - \frac{1\cdot3\cdot5}{9!}x^{-8} - \cdots\right)$$
$$+ B\left(x - \frac{1}{x}\right).$$

8. $x^2 \dfrac{d^2y}{dx^2} + (x + 2x^2) \dfrac{dy}{dx} - 4y = 0$,

$$y = Ax^2\left(1 - \frac{2\cdot2}{5}x + \frac{3\cdot2^2}{5\cdot6}x^2 - \frac{4\cdot2^3}{5\cdot6\cdot7}x^3 + \cdots\right) + B\left(\frac{1}{x^2} - \frac{4}{3x} + \frac{2}{3}\right).$$

9. $(x - x^2)\dfrac{d^2y}{dx^2} + 3\dfrac{dy}{dx} + 2y = 0$,

$$y = A(6 - 4x + x^2) + B\frac{1 - 4x}{x^2}.$$

Show also that $x^{-2}(1 - x)^4$ is an integral.

10. $(4x^3 - 14x^2 - 2x)\dfrac{d^2y}{dx^2} - (6x^2 - 7x + 1)\dfrac{dy}{dx}$
$$+ (6x - 1)y = 0,$$
$$y = Ax^{\frac{1}{2}}(1 + 2x) + B(1 - x).$$

11. $x^2 \dfrac{d^2y}{dx^2} + x^2 \dfrac{dy}{dx} + (x - 2)y = 0$,

$$y = \frac{A}{x} + Bx^2\left(\frac{1}{3} - \frac{1}{4}\frac{x}{1} + \frac{1}{5}\frac{x^2}{2!} - \frac{1}{6}\frac{x^3}{3!} + \cdots\right).$$

12. Denoting the integral in Ex. 11 by $Ay_1 + By_2$, find, by the method of Art. 146, an independent integral, and express the relation between the integrals.
$$y_2' = e^{-x}\left(\frac{2}{x} + 2 + x\right) = 2y_1 - y_2.$$

13. $x^2 \dfrac{d^2y}{dx^2} - x^2 \dfrac{dy}{dx} + (x - 2)y = 0$,

$$y = Ax^2\left(1 + \frac{x}{4} + \frac{x^2}{4\cdot5} + \frac{x^3}{4\cdot5\cdot6} + \cdots\right) + B\left(\frac{1}{x} + 1 + \frac{x}{2}\right).$$

Show also that $x^{-1}e^x$ is an integral.

14. $x^2(1-4x)\dfrac{d^2y}{dx^2} + [(1-n)x - (6-4n)x^2]\dfrac{dy}{dx}$
$\qquad\qquad\qquad + n(1-n)xy = 0,$

$y = Ax^n\left(1 + nx + \dfrac{n(n+3)}{2!}x^2 + \dfrac{n(n+4)(n+5)}{3!}x^3 + \ldots\right)$
$\qquad + B\left(1 - nx + \dfrac{n(n-3)}{2!}x^2 - \dfrac{n(n-4)(n-5)}{3!}x^3 + \ldots\right).$

15. $x^2\dfrac{d^2y}{dx^2} + (x + x^2)\dfrac{dy}{dx} + (x-9)y = 0,$

$y = Ax^{-3}\left(1 - \dfrac{2}{5}x + \dfrac{x^2}{20}\right) + Bx^3\left(1 - \dfrac{4}{1.7}x + \dfrac{4.5}{1.2.7.8}x^2 - \ldots\right).$

16. $(a^2 + x^2)\dfrac{d^2y}{dx^2} + x\dfrac{dy}{dx} - n^2 y = 0,$

$y = A\left(1 + \dfrac{n^2}{2!}\dfrac{x^2}{a^2} + \dfrac{n^2(n^2-4)}{4!}\dfrac{x^4}{a^4} + \ldots\right)$
$\qquad + Bx\left(1 + \dfrac{n^2-1}{3!}\dfrac{x^2}{a^2} + \dfrac{(n^2-1)(n^2-9)}{5!}\dfrac{x^4}{a^4} + \ldots\right).$

17. Denoting the integral given in Ex. 16 by $Ay_1 + By_2$, show that
$$[x + \sqrt{(a^2 + x^2)}]^n = a^n y_1 + na^{n-1} y_2,$$
and find the corresponding result when $n = 0$.

$\log[x + \sqrt{(a^2 + x^2)}] = \log a + \dfrac{x}{a} - \dfrac{1}{2}\dfrac{x^3}{3a^3} + \dfrac{1.3}{2.4}\dfrac{x^5}{5a^5} - \ldots$

18. Expand $\sin(a\sin^{-1}x)$ and $\cos(a\cos^{-1}x)$ by means of the differential equation
$$(1-x^2)\dfrac{d^2y}{dx^2} - x\dfrac{dy}{dx} + a^2 y = 0,$$
of which they are independent integrals.

$\sin(a\sin^{-1}x) = ax\left(1 - \dfrac{a^2-1}{3!}x^2 + \dfrac{(a^2-1)(a^2-9)}{5!}x^4 - \ldots\right),$

$\cos(a\sin^{-1}x) = 1 - \dfrac{a^2}{2!}x^2 + \dfrac{a^2(a^2-4)}{4!}x^4 - \ldots$

XV.

Case of Equal Values of m.

171. If the two roots of the equation determining m are equal, we can determine one integral of the form $y = \Sigma A_r x^{m+rs}$ by the process given in the foregoing articles; but there is no other integral of this form. We therefore require an independent integral of some other form.

For example, let the given equation be

$$x(1 - x^2)\frac{d^2y}{dx^2} + (1 - 3x^2)\frac{dy}{dx} - xy = 0, \quad \ldots \quad (1)$$

a binomial equation, in which we may take $s = 2$, or $s = -2$. Assuming

$$y = \Sigma_0^\infty A_r x^{m+2r},$$

we have, by substitution,

$$\Sigma_0^\infty [(m + 2r)^2 A_r x^{m+2r-1} - (m + 2r + 1)^2 A_r x^{m+2r+1}] = 0.$$

Equating to zero the coefficient of x^{m+2r-1}, we have

$$(m + 2r)^2 A_r - (m + 2r - 1)^2 A_{r-1} = 0. \quad \ldots \quad (2)$$

Putting $r = 0$, $m^2 A_0 = 0$; whence

$$m = 0,$$

the two values of m being identical. Putting $m = 0$ in equation (2), the relation between consecutive coefficients is

$$A_r = \frac{(2r - 1)^2}{(2r)^2} A_{r-1};$$

whence we find the integral

$$A_0 y_1 = A_0\left(1 + \frac{1^2}{2^2}x^2 + \frac{1^2 \cdot 3^2}{2^2 \cdot 4^2}x^4 + \frac{1^2 \cdot 3^2 \cdot 5^2}{2^2 \cdot 4^2 \cdot 6^2}x^6 + \ldots\right). \quad (3)$$

172. To obtain a new integral, we shall first suppose the given equation to be so modified that one of the equal factors in the first term of equation (2) is changed to $m + 2r - h$, so that one of the values of m becomes equal to h, while the other value remains equal to zero. We shall then obtain the complete integral of the modified equation, in which, after some transformation, we shall put $h = 0$, and thus obtain the complete integral of equation (1).

The altered relation between consecutive coefficients may be written

$$A_r = \frac{(m + 2r - 1)^2}{(m + 2r)(m + 2r - h')} A_{r-1}, \quad \ldots \quad (4)$$

in which, for a reason which will presently be explained, h' is put in the place of h. Hence, when $m = 0$, we have

$$A_r = \frac{(2r - 1)^2}{2r(2r - h')} A_{r-1};$$

and the first integral now is

$$y_1 = 1 + \frac{1^2}{2(2 - h')} x^2 + \frac{1^2 \cdot 3^2}{2 \cdot 4(2 - h')(4 - h')} x^4 + \ldots \quad (5)$$

Putting $m = h$ in equation (4), we have

$$B_r = \frac{(2r - 1 + h)^2}{(2r + h)(2r - h' + h)} B_{r-1};$$

and the second integral is

$$y_2 = x^h \left(1 + \frac{(1 + h)^2}{(2 + h)(2 - h' + h)} x^2 \right.$$

$$\left. + \frac{(1 + h)^2 (3 + h)^2}{(2 + h)(4 + h)(2 - h' + h)(4 - h' + h)} x^4 + \ldots \right). \quad (6)$$

§ XV.]   CASE OF EQUAL VALUES OF m.   183

The object of introducing h' in equation (4), in place of the equal quantity h, is that, when equation (6) is written in the form

$$y_2 = x^h \psi(h),$$

$\psi(h)$ shall be such a function of h that, by equation (5),

$$y_1 = \psi(0).$$

Developing y_2 in powers of h, we have, since $x^h = e^{h \log x}$,

$$y_2 = (1 + h \log x + \ldots)[y_1 + h\psi'(0) + \ldots];$$

hence the complete integral is

$$y = A_0 y_1 + B_0 y_1 + B_0 h [y_1 \log x + \psi'(0) + \ldots];$$

or, replacing the constants $A_0 + B_0$ and $B_0 h$ by A and B,

$$y = Ay_1 + By_1 \log x + B\psi'(0) + \ldots, \quad \ldots \quad (7)$$

in which we have retained all the terms which do not vanish with h, and, when $h = 0$, y_1 resumes the value given in equation (3).

173. It remains to express $\psi'(0)$ in terms of x. In doing this, we may, since h' is finally to be put equal to zero, make this substitution in the value of $\psi(h)$ at once, and write

$$\psi(h) = 1 + \frac{(1 + h)^2}{(2 + h)^2} x^2 + \frac{(1 + h)^2 (3 + h)^2}{(2 + h)^2 (4 + h)^2} x^4 + \ldots \quad (8)$$

Denote the coefficient of x^{2r} in this series by H_r, so that $H_0 = 1$, and when $r > 0$,

$$H_r = \frac{(1 + h)^2 (3 + h)^2 \ldots (2r - 1 + h)^2}{(2 + h)^2 (4 + h)^2 \ldots (2r + h)^2}; \quad \ldots \quad (9)$$

then
$$\psi(h) = \Sigma_0^\infty H_r x^{2r};$$
and
$$\psi'(h) = \Sigma_1^\infty \frac{dH_r}{dh} x^{2r} = \Sigma_1^\infty H_r \frac{d \log H_r}{dh} x^{2r}, \quad \ldots \quad (10)$$

in which unity is taken as the lower limit because $\dfrac{dH_0}{dh} = 0$. From equation (9),

$$\frac{d \log H_r}{dh} = \frac{2}{1+h} + \frac{2}{3+h} + \cdots + \frac{2}{2r-1+h}$$
$$- \frac{2}{2+h} - \frac{2}{4+h} - \cdots - \frac{2}{2r+h},$$

which, when $h = 0$, becomes

$$\left. \frac{d \log H_r}{dh} \right]_0 = \frac{2}{1} + \frac{2}{3} + \cdots + \frac{2}{2r-1} - \frac{2}{2} - \frac{2}{4} - \cdots - \frac{2}{2r};$$

whence, putting $h = 0$ in equation (10), and denoting $\psi'(0)$, when thus expressed as a series in x, by y',

$$y' = \frac{1^2}{2^2}\left(\frac{2}{1} - \frac{2}{2}\right) x^2 + \frac{1^2 \cdot 3^2}{2^2 \cdot 4^2}\left(\frac{2}{1} + \frac{2}{3} - \frac{2}{2} - \frac{2}{4}\right) x^4 + \cdots \quad (11)$$

Hence, when $h = 0$, equation (7) gives for the complete integral of equation (1)*

$$y = Ay_1 + B(y_1 \log x + y'),$$

where y_1 and y' are defined by equations (3) and (11).

---

* For the complete integral when we take $s = -2$, see Ex. XV. 7.

§ XV.]   *INTEGRALS OF THE LOGARITHMIC FORM.*   185

Case in which the Values of m differ by a Multiple of s.

174. When the two values of m differ by a multiple of s, the initial term of one of the series will appear as a term of the other series; and the coefficient of this term will contain a zero factor in its denominator. Hence, unless a zero factor occurs in the numerator,* the coefficient will be infinite; and, as in the preceding case, it is impossible to obtain two independent integrals of the form $\Sigma A_r x^{m+rs}$. For example, let the given equation be

$$x^2(1+x)\frac{d^2y}{dx^2} + x\frac{dy}{dx} + (1-2x)y = 0. \quad \ldots \quad (1)$$

Putting $y = A_0 x^m$ in the first member, the result is

$$A_0(m^2+1)x^m + A_0(m^2-m-2)x^{m+1}.$$

Choosing the second term as that which is to vanish by the determination of m, because the first would give imaginary roots, we have

$$m = -1 \quad \text{or} \quad m = 2, \quad \text{and} \quad s = -1;$$

hence, putting $y = \Sigma_0^\infty A_r x^{m-r}$,

$$\Sigma_0^\infty \{(m-r+1)(m-r-2)A_r x^{m-r+1}$$
$$+ [(m-r)^2+1]A_r x^{m-r}\} = 0;$$

and, equating to zero the coefficient of x^{m-r+1},

$$(m-r+1)(m-r-2)A_r + [(m-r+1)^2+1]A_{r-1} = 0. \quad (2)$$

---

* It is immaterial whether the zero factor in the numerator first occurs in the term in question, or in a preceding term; the result is a finite solution. An example of this exceptional case has already occurred in Art. 168, where $s = 1$, and the values of m differ by an integer.

When $m = -1$, the relation between consecutive coefficients is

$$A_r = -\frac{r^2+1}{r(r+3)}A_{r-1};$$

and the first integral is

$$A_0 y_1 = A_0 x^{-1}\left(1 - \frac{2}{1.4}x^{-1} + \frac{2.5}{1.2.4.5}x^{-2} - \frac{2.5.10}{1.2.3.4.5.6}x^{-3} + \ldots\right). \quad (3)$$

Putting $m = 2$, the relation is

$$B_r = -\frac{(r-3)^2+1}{(r-3)r}B_{r-1};$$

and the second integral takes the form

$$B_0 y_2 = B_0 x^2\left(1 - \frac{5}{-2.1}x^{-1} + \frac{5.2}{-2(-1).1.2}x^{-2} - \frac{5.2.1}{-2(-1).0.1.2.3}x^{-3} + \ldots\right), \quad (4)$$

in which the coefficient of x^{-1} is infinite. Thus, the second integral of the form $\Sigma A_r x^{m+rs}$ fails, and we require an independent integral of some other form.

175. To obtain the new integral, we proceed as in Art. 172. Thus, supposing the second factor in the first term of equation (2) to be changed to $m - r - 2 - h$, so that the second value of m is now $2 + h$ instead of 2, and using h' as in Art. 172, the relation between consecutive coefficients now is

$$A_r = -\frac{(m-r+1)^2+1}{(m-r+1)(m-r-2-h')}A_{r-1}. \quad (5)$$

§ XV.] *INTEGRALS OF THE LOGARITHMIC FORM.* 187

When $m = -1$, this becomes

$$A_r = -\frac{r^2 + 1}{r(r + 3 + h')} A_{r-1},$$

and we have

$$A_0 y_1 = A_0 x^{-1}\left(1 - \frac{2}{1(4 + h')}x^{-1}\right.$$
$$\left. + \frac{2 \cdot 5}{1 \cdot 2(4 + h')(5 + h')}x^{-2} - \ldots\right). \quad (6)$$

Putting $m = 2 + h$, the relation between the coefficients in y_2

$$B_r = -\frac{(r - 3 - h)^2 + 1}{(r - 3 - h)(r + h' - h)} B_{r-1};$$

and the new value of $B_0 y_2$ is

$$B_0 y_2 = B_0 x^{2+h}\left(1 - \frac{(-2 - h)^2 + 1}{(-2 - h)(1 + h' - h)}x^{-1} + \ldots\right),$$

in which the first term which becomes infinite when $h = 0$ is

$$-B_0 \frac{[(-2-h)^2 + 1][(-1-h)^2 + 1][(-h)^2 + 1]}{(-2-h)(-1-h)(-h)(1+h'-h)(2+h'-h)(3+h'-h)} x^{-1+h}. \quad (7)$$

Denoting the coefficient of this term by $\dfrac{B}{h}$, and the sum of the preceding terms in y_2 by T, we may write

$$B_0 y_2 = B_0 T$$
$$+ \frac{B}{h}x^{-1+h}\left(1 - \frac{(1 - h)^2 + 1}{(1 - h)(4 + h' - h)}x^{-1} + \ldots\right). \quad (8)$$

If now we write this equation in the form

$$B_0 y_2 = B_0 T + \frac{B}{h}x^h \psi(h),$$

equation (6) shows that $y_1 = \psi(0)$; hence the complete integral may be written

$$y = A_0 y_1 + B_0 T + \frac{B}{h}(1 + h \log x + \ldots)[y_1 + h\psi'(0) + \ldots],$$

or, putting A for the constant $A_0 + \frac{B}{h}$,

$$y = A y_1 + B_0 T + B y_1 \log x + B \psi'(0) + \ldots \ldots \quad (9)$$

In this equation we have retained all the terms which do not vanish with h; from the value of B, as defined by the expression (7), we see that, when $h = 0$,

$$B = B_0 \frac{5 \cdot 2 \cdot 1}{(-2)(-1) \cdot 1 \cdot 2 \cdot 3} = \frac{5}{6} B_0; \ldots \quad (10)$$

and, when $h = 0$, we have, from equation (4),

$$T = x^2 + \tfrac{5}{2} x + \tfrac{5}{6}. \ldots \ldots \quad (11)$$

176. The expression for $\psi'(0)$ as a series in x, which we shall denote by y', is found exactly as in Art. 173. Putting $h' = 0$ at once, in the value of $\psi(h)$ as defined by equation (8), we have

$$\psi(h) = x^{-1}\left(1 - \frac{(1-h)^2 + 1}{(1-h)(4-h)} x^{-1} \right.$$
$$\left. + \frac{[(1-h)^2 + 1][(2-h)^2 + 1]}{(1-h)(2-h)(4-h)(5-h)} x^{-2} - \ldots \right);$$

and, writing this in the form

$$\psi(h) = x^{-1} \Sigma_0^\infty H_r (-x)^{-r},$$

we have $H_0 = 1$, and, when $r > 1$,

$$H_r = \frac{[(1-h)^2 + 1][(2-h)^2 + 1] \ldots [(r-h)^2 + 1]}{(1-h)(2-h)\ldots(r-h)(4-h)\ldots(r+3-h)}.$$

§ XV.] INTEGRALS OF THE LOGARITHMIC FORM.

Hence

$$\psi'(h) = x^{-1} \sum_{1}^{\infty} (-1)^r H_r \frac{d \log H_r}{dh} x^{-r}, \quad \ldots \quad (12)$$

in which

$$\frac{d \log H_r}{dh} = -\frac{2(1-h)}{(1-h)^2 + 1} - \frac{2(2-h)}{(2-h)^2 + 1} - \ldots$$

$$- \frac{2(r-h)}{(r-h)^2 + 1} + \frac{1}{1-h} + \frac{1}{2-h} + \ldots$$

$$+ \frac{1}{r-h} + \frac{1}{4-h} + \ldots + \frac{1}{r+3-h}.$$

When $h = 0$, this becomes

$$\left[\frac{d \log H_r}{dh}\right]_0 = -\frac{2}{1^2+1} - \frac{4}{2^2+1} - \ldots - \frac{2r}{r^2+1}$$

$$+ \frac{1}{1} + \frac{1}{2} + \ldots + \frac{1}{r} + \frac{1}{4} + \frac{1}{5} + \ldots + \frac{1}{r+3};$$

hence, putting $h = 0$ in equation (12), we have

$$y' = x^{-1}\left[\frac{2}{1 \cdot 4}\left(\frac{2}{2} - \frac{1}{1} - \frac{1}{4}\right)x^{-1}\right.$$

$$\left. - \frac{2 \cdot 5}{1 \cdot 2 \cdot 4 \cdot 5}\left(\frac{2}{2} + \frac{4}{5} - \frac{1}{1} - \frac{1}{2} - \frac{1}{4} - \frac{1}{5}\right)x^{-2} + \ldots\right]. \quad (13)$$

Now, putting $h = 0$ in equation (9), substituting $B_0 = \tfrac{8}{6}B$ from equation (10) and the value of T from equation (11), we have, for the complete integral of equation (1),

$$y = Ay_1 + B(\tfrac{8}{6}x^2 + 3x + 3 + y_1 \log x + y'),$$

where y_1 and y' are defined by equations (3) and (13).

Special Forms of the Particular Integral.

177. We have seen, in Art. 166, that the particular integral when the second member of the given equation is a power of x, may be expressed in the form of a series similar to those which constitute the complementary function. Special cases arise in which the particular integral either admits of expression as a finite series, or can only be expressed in the logarithmic form considered in the preceding articles. In illustration, let us take the equation

$$(1 - x^2)\frac{d^2y}{dx^2} - x\frac{dy}{dx} = px^a, \quad \ldots \quad (1)$$

of which the complementary function is $A \sin^{-1}x + B$. Putting $y = \Sigma_0^\infty A_r x^{m+2r}$, we have

$$\Sigma_0^\infty A_r[(m + 2r)(m + 2r - 1)x^{m+2r-2} - (m + 2r)^2 x^{m+2r}] = px^a; \quad (2)$$

whence, when $r > 0$,

$$(m + 2r)(m + 2r - 1)A_r - (m + 2r - 2)^2 A_{r-1} = 0,$$

and the relation between consecutive coefficients is

$$A_r = \frac{(m + 2r - 2)^2}{(m + 2r)(m + 2r - 1)} A_{r-1}. \quad \ldots \quad (3)$$

For the complementary function, we have $m = 1$, or $m = 0$. Putting $m = 1$ in equation (3),

$$A_r = \frac{(2r - 1)^2}{2r(2r + 1)} A_{r-1};$$

whence

$$y_1 = x\left(1 + \frac{1^2}{2.3}x^2 + \frac{1^2 \cdot 3^2}{2.3.4.5}x^4 + \ldots\right) \quad \ldots \quad (4)$$

§ XV.] SPECIAL FORMS OF THE PARTICULAR INTEGRAL.

This is the value of $\sin^{-1}x$. The series corresponding to $m = 0$ reduces to a single term, so that

$$y_2 = 1.$$

For the particular integral Y, we have, from equation (2),

$$m(m - 1)A_0 x^{m-2} = px^a;$$

whence

$$m = a + 2, \quad \text{and} \quad A_0 = \frac{p}{(a + 1)(a + 2)}.$$

Putting $m = a + 2$ in the relation (3),

$$A_r = \frac{(a + 2r)^2}{(a + 2r + 1)(a + 2r + 2)} A_{r-1};$$

hence

$$Y = \frac{px^{a+2}}{(a + 1)(a + 2)}\left(1 + \frac{(a + 2)^2}{(a + 3)(a + 4)}x^2 \right.$$
$$\left. + \frac{(a + 2)^2(a + 4)^2}{(a + 3)(a + 4)(a + 5)(a + 6)}x^4 + \ldots\right). \quad (5)$$

This equation gives the particular integral except when a is a negative integer; for instance, when $a = 0$, and $p = 2$, it gives

$$Y = x^2\left(1 + \frac{2^2}{3.4}x^2 + \frac{2^2 \cdot 4^2}{3.4.5.6}x^4 + \ldots\right),$$

which, as will be found by comparing the finite solution of equation (1) in the case considered, is the value of $(\sin^{-1}x)^2$.

178. Now, in the first place, if a is a positive odd integer, all the powers of x which occur in Y occur also in y_1; and, when this is the case, we can obtain a particular integral in the form of a finite series. For example, if $a = 3$, we have

$$Y = \frac{px^5}{4.5}\left(1 + \frac{5^2}{6.7}x^2 + \frac{5^2 \cdot 7^2}{6.7.8.9}x^4 + \ldots\right).$$

If we write this equation in the form

$$\frac{1^2 \cdot 3^2}{2 \cdot 3 p} Y = \frac{1^2 \cdot 3^2 x^5}{2 \cdot 3 \cdot 4 \cdot 5}\left(1 + \frac{5^2}{6 \cdot 7} x^2 + \cdots\right),$$

the second member is equivalent to the series y_1, equation (4), with the exception of its first two terms. Thus

$$\frac{3Y}{2p} = y_1 - \left(x + \frac{1}{6} x^3\right), \quad \text{or} \quad Y = \frac{2p}{3} y_1 - \frac{2p}{3}\left(x + \frac{1}{6} x^3\right);$$

and, since the first term of this expression is included in the complementary function, we have the particular integral

$$Y = -\frac{2p}{3} x - \frac{p}{9} x^3.$$

This finite particular integral would have been found directly had we employed a series in descending powers of x.

179. In the next place, when a is a negative odd integer, the initial term of y_1 will occur in Y with an infinite coefficient. Thus, if $a = -3$ in equation (5), Art. 177, the second term contains the first power of x and has an infinite coefficient. To obtain the particular integral in this case, suppose first that $a = -3 + h$; then equation (5) gives

$$Y = \frac{p x^{-1+h}}{(-2+h)(-1+h)}$$

$$+ \frac{p(-1+h)^2 x^{1+h}}{(-2+h)(-1+h)h(1+h)}\left(1 + \frac{(1+h)^2}{(2+h)(3+h)} x^2 + \cdots\right).$$

Putting

$$\psi(h) = x\left(1 + \frac{(1+h)^2}{(2+h)(3+h)} x^2 + \cdots\right),$$

§ XV.] SPECIAL FORMS OF THE PARTICULAR INTEGRAL. 193

equation (4), Art. 177, shows that $y_1 = \psi(0)$; and we may write

$$Y = T + \frac{N}{h}(1 + h\log x + \ldots)[y_1 + h\psi'(0) + \ldots]$$

here N is a quantity which remains finite when $h = 0$. Expanding, and rejecting the term $\frac{N}{h}y_1$, which is included in the complementary function, we may now take, for the particular integral,

$$Y' = T + Ny_1 \log x + N\psi'(0) + \ldots,$$

in which we have retained all the terms which do not vanish with h. When $h = 0$, the values of T and N are $\frac{p}{2x}$ and $\frac{p}{2}$ respectively; and, finding the value of $\psi'(0)$, as in Arts. 173 and 176, we have, for the particular integral,

$$\frac{p}{2x} + \frac{p}{2}\sin^{-1}x . \log x + \frac{px}{2}\left[\frac{1^2}{2.3}\left(\frac{2}{1} - \frac{1}{2} - \frac{1}{3}\right)x^2\right.$$
$$\left. + \frac{1^2 . 3^2}{2.3.4.5}\left(\frac{2}{1} + \frac{2}{3} - \frac{1}{2} - \frac{1}{3} - \frac{1}{4} - \frac{1}{5}\right)x^4 + \ldots\right].$$

180. In like manner, when a is a negative even integer, the term containing x^0, corresponding to y_2, occurs in Y with an infinite coefficient. Thus, if $a = -4$, the second term of the series in equation (5), Art. 177, is infinite. But, putting $a = -4 + h$, we have

$$V = \frac{px^{-2+h}}{(-3+h)(-2+h)} + \frac{p(-2+h)^2}{(-3+h)(-2+h)(-1+h)h}$$
$$\times (1 + h\log x + \ldots)\left(1 + \frac{h^2}{(1+h)(2+h)}x^2 + \ldots\right),$$

or

$$Y = T + \frac{N}{h}(1 + h\log x + \ldots)\psi(h).$$

In this case, when $\psi(h)$ is expanded in powers of h, the first term is unity, and there is no term containing the first power of h; hence, rejecting the term $\dfrac{N}{h}$ which is included in the complementary function, and then putting $h = 0$, we have the particular integral

$$Y' = \frac{p}{6x^2} - \frac{2p}{3}\log x.$$

Examples XV.

Integrate in series the following differential equations:—

1. $x\dfrac{d^2y}{dx^2} + \dfrac{dy}{dx} + y = 0,$

$y = (A + B\log x)\left(1 - \dfrac{x}{1^2} + \dfrac{x^2}{1^2.2^2} - \dfrac{x^3}{1^2.2^2.3^2} + \cdots\right)$

$\qquad + 2B\left[\dfrac{x}{1^2} - \dfrac{x^2}{1^2.2^2}\left(1 + \dfrac{1}{2}\right) + \dfrac{x^3}{1^2.2^2.3^2}\left(1 + \dfrac{1}{2} + \dfrac{1}{3}\right) - \cdots\right].$

2. $x\dfrac{d^2y}{dx^2} + \dfrac{dy}{dx} + pxy = 0,$

$y = (A + B\log x)\left(1 - \dfrac{px^2}{2^2} + \dfrac{p^2x^4}{2^2.4^2} - \cdots\right)$

$\qquad + B\left[\dfrac{px^2}{2^2} - \dfrac{p^2x^4}{2^2.4^2}\left(1 + \dfrac{1}{2}\right) + \dfrac{p^3x^6}{2^2.4^2.6^2}\left(1 + \dfrac{1}{2} + \dfrac{1}{3}\right) - \cdots\right].$

3. $x\dfrac{d^2y}{dx^2} + y = 0,$

$y = (A + B\log x)x\left(1 - \dfrac{x}{1.2} + \dfrac{x^2}{1.2^2.3} - \dfrac{x^3}{1.2^2.3^2.4} + \cdots\right)$

$\qquad -B + Bx\left[\dfrac{x}{1.2}\dfrac{3}{1.2} - \dfrac{x^2}{1.2^2.3}\left(\dfrac{3}{1.2} + \dfrac{5}{2.3}\right) + \cdots\right].$

EXAMPLES. [§ XV.]

4. $x^3\dfrac{d^2y}{dx^2} - (2x - 1)y = 0$,

$y = (A + B\log x)x^{-1}\left(1 - \dfrac{x^{-1}}{1.4} + \dfrac{x^{-2}}{1.2.4.5} - \dfrac{x^{-3}}{1.2.3.4.5.6} + \cdots\right)$
$+ 3B(4x^2 + 2x + 1)$
$- Bx^{-1}\left[\dfrac{x^{-1}}{1.4}\left(1 + \dfrac{1}{4}\right) - \dfrac{x^{-2}}{1.2.4.5}\left(1 + \dfrac{1}{2} + \dfrac{1}{4} + \dfrac{1}{5}\right) + \cdots\right]$.

5. $x^2\dfrac{d^2y}{dx^2} + x(x + 1)\dfrac{dy}{dx} + (3x - 1)y = 0$,

$y = (A + B\log x)x\left(1 - \dfrac{4}{3}\dfrac{x}{1} + \dfrac{5}{3}\dfrac{x^2}{2!} - \dfrac{6}{3}\dfrac{x^3}{3!} + \cdots\right) - \dfrac{B}{3}(x^{-1} + 2)$
$+ Bx\left[\dfrac{4}{3}\dfrac{x}{1}\left(\dfrac{1}{3} - \dfrac{1}{4} + 1\right) - \dfrac{5}{3}\dfrac{x^2}{2!}\left(\dfrac{1}{3} - \dfrac{1}{5} + 1 + \dfrac{1}{2}\right) + \cdots\right]$.

6. $(x - x^2)\dfrac{d^2y}{dx^2} - y = 0$,

$y = (A + B\log x)x\left(1 + \dfrac{1}{1.2}x + \dfrac{1.3}{1.2^2.3}x^2 + \dfrac{1.3.7}{1.2^2.3^2.4}x^3 + \cdots\right)$
$+ B + Bx\left[\dfrac{1}{1.2}\left(\dfrac{1}{1} - \dfrac{1}{1} - \dfrac{1}{2}\right)x + \dfrac{1.3}{1.2^2.3}\left(\dfrac{1}{1} + \dfrac{3}{3} - \dfrac{1}{1} - \dfrac{2}{2} - \dfrac{1}{3}\right)x^2\right.$
$\left. + \dfrac{1.3.7}{1.2^2.3^2.4}\left(\dfrac{1}{1} + \dfrac{3}{3} + \dfrac{5}{7} - \dfrac{1}{1} - \dfrac{2}{2} - \dfrac{2}{3} - \dfrac{1}{4}\right)x^3 + \cdots\right]$.

7. Find the integral of

$$x(1 - x^2)\dfrac{d^2y}{dx^2} + (1 - 3x^2)\dfrac{dy}{dx} - xy = 0,$$

[equation (1), Art. 171,] when $x > 1$.

$y = (A + B\log x)x^{-1}\left(1 + \dfrac{1^2}{2^2}x^{-2} + \dfrac{1^2.3^2}{2^2.4^2}x^{-4} + \cdots\right)$
$- 2Bx^{-1}\left[\dfrac{1^2}{2^2}\dfrac{1}{1.2}x^{-2} + \dfrac{1^2.3^2}{2^2.4^2}\left(\dfrac{1}{1.2} + \dfrac{1}{3.4}\right)x^{-4} + \cdots\right]$.

8. $\frac{d^2y}{dx^2} + \frac{ay}{x^{\frac{3}{2}}} = 0,$

$y = (A + B \log x)x\left(1 - \frac{4ax^{\frac{1}{2}}}{1.3} + \frac{4^2a^2x}{1.2.3.4} - \frac{4^3a^3x^{\frac{3}{2}}}{1.2.3^2.4.5} + \ldots\right)$

$\quad - \frac{B}{4a^2}(1 + 4ax^{\frac{1}{2}})$

$\quad + 2Bx\left[\frac{4ax^{\frac{1}{2}}}{1.3}\left(\frac{1}{1} + \frac{1}{3}\right) - \frac{4^2a^2x}{1.2.3.4}\left(\frac{1}{1} + \frac{1}{2} + \frac{1}{3} + \frac{1}{4}\right) + \ldots\right].$

9. $x^3\frac{d^2y}{dx^2} - (x^2 + 4x)\frac{dy}{dx} + 4y = 0,$

$y = Ax^4e^x + B(2x - x^2 + x^3 + x^4e^x \log x)$

$\quad - Bx^4\left[x + \frac{x^2}{2!}\left(1 + \frac{1}{2}\right) + \frac{x^3}{3!}\left(1 + \frac{1}{2} + \frac{1}{3}\right) + \ldots\right].$

10. $x(1 - x^2)\frac{d^2y}{dx^2} + (1 - x^2)\frac{dy}{dx} + xy = 0,$

$y = (A + B \log x)\left(1 - \frac{1}{2^2}x^2 - \frac{1.3}{2^2.4^2}x^4 - \frac{1.3^2.5}{2^2.4^2.6^2}x^6 - \ldots\right)$

$\quad + B\left[\frac{1}{2^2}x^2 + \frac{1.3}{2^2.4^2}\left(\frac{2}{4} - \frac{1}{3}\right)x^4\right.$

$\quad\quad \left. + \frac{1.3^2.5}{2^2.4^2.6^2}\left(\frac{2}{4} + \frac{2}{6} - \frac{2}{3} - \frac{1}{5}\right)x^6 + \ldots\right].$

11. $4x(1 - x)\frac{d^2y}{dx^2} - 4\frac{dy}{dx} - y = 0,$

$y = (A + B \log x)x^2\left(1 + \frac{3^2}{2.6}x + \frac{3^2.5^2}{2.4.6.8}x^2 + \frac{3^2.5^2.7^2}{2.4.6^2.8.10}x^3 + \ldots\right)$

$\quad - B(32 - 8x) + 2Bx^2\left[\frac{3^2}{2.6}\left(\frac{2}{3} - \frac{1}{2} - \frac{1}{6}\right)x\right.$

$\quad\quad \left. + \frac{3^2.5^2}{2.4.6.8}\left(\frac{2}{3} - \frac{1}{2} - \frac{1}{6} + \frac{2}{5} - \frac{1}{4} - \frac{1}{8}\right)x^2 + \ldots\right].$

§ XV.]   EXAMPLES.   197

12. $x^3 \dfrac{d^2y}{dx^2} + y = x^{\frac{1}{2}}$,

$$y = (A + B\log x)\left(1 - \dfrac{x^{-1}}{1.2} + \dfrac{x^{-2}}{1.2^2.3} - \dfrac{x^{-3}}{1.2^2.3^2.4} + \cdots\right)$$

$$+ Bx - B\left[\dfrac{x^{-1}}{1.2}\left(\dfrac{1}{1} + \dfrac{1}{2}\right) - \dfrac{x^{-2}}{1.2^2.3}\left(\dfrac{1}{1} + \dfrac{2}{2} + \dfrac{1}{3}\right) + \cdots\right]$$

$$- 4x^{\frac{1}{2}}\left(1 - \dfrac{4x^{-1}}{1.3} + \dfrac{4^2 x^{-2}}{1.3^2.5} - \dfrac{4^3 x^{-3}}{1.3^2.5^2.7} + \cdots\right).$$

13. $2x^2 \dfrac{d^2y}{dx^2} - (3x + 2)\dfrac{dy}{dx} + \dfrac{2x - 1}{x} y = x^{\frac{1}{2}}$,

$$y = Ax^2\left(1 - \dfrac{5}{1}\dfrac{x^{-1}}{1!} + \dfrac{5.3}{1.-1}\dfrac{x^{-2}}{2!} - \dfrac{5.3.1}{1.-1.-3}\dfrac{x^{-3}}{3!} + \cdots\right)$$

$$+ \left(B - \dfrac{1}{3}\log x\right)\left(x^{\frac{1}{2}} + \dfrac{2}{5}x^{-\frac{1}{2}}\right) - \dfrac{8}{25}x^{-\frac{1}{2}}$$

$$- \dfrac{2}{105}x^{-\frac{3}{2}}\left(1 - \dfrac{2}{3.9}x^{-1} + \dfrac{2.4}{3.4.9.11}x^{-2} - \cdots\right).$$

14. Express the particular integral of the equation

$$(x - x^2)\dfrac{d^2y}{dx^2} + 3\dfrac{dy}{dx} + 2y = 3x^2,$$

(α) in the form of an ascending series; (β) in the form of a descending series; (γ) as a finite expression.   [See Example XIV. 9, for the complementary function.]

(α) $Y = \dfrac{x^3}{5}\left(1 + \dfrac{1}{6}x + \dfrac{1.2}{6.7}x^2 + \cdots\right).$

(β) $Y = -\dfrac{(x-1)^4}{x^2}\log x + 5x - \dfrac{25}{2} + \dfrac{35}{3}x^{-1} - \dfrac{25}{6}x^{-2}$

$$- \dfrac{x^{-3}}{5}\left(1 + \dfrac{1}{6}x^{-1} + \dfrac{1.2}{6.7}x^{-2} + \cdots\right).$$

(γ) $Y = \dfrac{(1-x)^4}{x^2}\displaystyle\int \dfrac{x^4 dx}{(1-x)^5}.$

CHAPTER VIII.

THE HYPERGEOMETRIC SERIES.

XVI.

General Solution of the Binomial Equation of the Second Order.

181. The symbol $F(\alpha, \beta, \gamma, z)$ is used to denote the series

$$1 + \frac{\alpha\beta}{1\cdot\gamma}z + \frac{\alpha(\alpha+1)\beta(\beta+1)}{1\cdot 2\cdot\gamma(\gamma+1)}z^2 + \frac{\alpha(\alpha+1)(\alpha+2)\beta(\beta+1)(\beta+2)}{1\cdot 2\cdot 3\cdot\gamma(\gamma+1)(\gamma+2)}z^3 + \cdots$$

which is known as the *hypergeometric series*. Regarding the first three elements, α, β, and γ, as constants, and the fourth as a variable containing x, the series includes a great variety of functions of x. In fact we shall now show that one, and generally both, of the independent integrals of a binomial differential equation of the second order whose second member is zero can be expressed by means of hypergeometric series in which the variable element is a power of x.

182. Using the notation of Art. 123,

$$x\frac{d}{dx} = \vartheta, \quad \text{whence} \quad x^2\frac{d^2}{dx^2} = \vartheta(\vartheta - 1),$$

we may, as in Art. 167 (first multiplying by a suitable power of x), reduce the binomial equation to the form

$$f(\vartheta)y + x^s\phi(\vartheta)y = 0, \quad \ldots \ldots \ldots \quad (1)$$

§ XVI.] BINOMIAL EQUATION OF THE SECOND ORDER. 199

in which f and ϕ are algebraic functions, one of which will be of a degree corresponding to the order of the equation, and the other of the same or an inferior degree. If the equation is of the second order, it may be written

$$(\vartheta - a)(\vartheta - b)y - qx^s(\vartheta - c)(\vartheta - d)y = 0, \quad . \quad . \quad (2)$$

in which q and s are positive or negative constants. Furthermore, the equation is readily reduced to a form in which q and s are each equal to unity; for, putting

$$z = qx^s, \quad \text{and} \quad \vartheta' = z\frac{d}{dz},$$

we have

$$\vartheta' = qx^s \frac{d}{qsx^{s-1}dx} = \frac{1}{s}\vartheta, \quad \text{or} \quad \vartheta = s\vartheta';$$

and, substituting, equation (2) becomes

$$\left(\vartheta' - \frac{a}{s}\right)\left(\vartheta' - \frac{b}{s}\right)y - z\left(\vartheta' - \frac{c}{s}\right)\left(\vartheta' - \frac{d}{s}\right)y = 0. \quad . \quad . \quad (3)$$

183. We may, therefore, suppose the binomial equation of the second order reduced to the standard form

$$(\vartheta - a)(\vartheta - b)y - x(\vartheta - c)(\vartheta - d)y = 0. \quad . \quad . \quad (1)$$

Substituting in this equation

$$y = \Sigma_0^\infty A_r x^{m+r},$$

we have

$$\Sigma_0^\infty A_r[(m+r-a)(m+r-b)x^{m+r} - (m+r-c)(m+r-d)x^{m+r+1}] = 0,$$

and, equating to zero the coefficient of x^{m+r},

$$(m+r-a)(m+r-b)A_r - (m+r-1-c)(m+r-1-d)A_{r-1} = 0.$$

This gives the relation between consecutive coefficients,

$$A_r = \frac{(m-c+r-1)(m-d+r-1)}{(m-a+r)(m-b+r)} A_{r-1},$$

and, when $r = 0$,

$$(m-a)(m-b)A_0 = 0;$$

whence $m = a$ or $m = b$. Putting $m = a$, we have for the first integral

$$y_1 = x^a \left(1 + \frac{(a-c)(a-d)}{1(a-b+1)} x \right.$$
$$\left. + \frac{(a-c)(a-c+1)(a-d)(a-d+1)}{1 \cdot 2(a-b+1)(a-b+2)} x^2 + \ldots \right), \quad (2)$$

and, interchanging a and b, the second integral is

$$y_2 = x^b \left(1 + \frac{(b-c)(b-d)}{1(b-a+1)} x \right.$$
$$\left. + \frac{(b-c)(b-c+1)(b-d)(b-d+1)}{1 \cdot 2(b-a+1)(b-a+2)} x^2 + \ldots \right). \quad (3)$$

Thus, putting

$$\left. \begin{array}{l} a - c = \alpha \\ a - d = \beta \\ a - b + 1 = \gamma \end{array} \right\}, \quad \ldots \ldots \ldots (4)$$

the first integral is

$$y_1 = x^a \left(1 + \frac{\alpha \beta}{1 \cdot \gamma} x + \frac{\alpha(\alpha+1)\beta(\beta+1)}{1 \cdot 2 \cdot \gamma(\gamma+1)} x^2 + \ldots \right)$$
$$= x^a F(\alpha, \beta, \gamma, x), \quad \ldots \ldots \ldots \ldots (5)$$

and the second may be written

$$y_2 = x^b F(\alpha', \beta', \gamma', x), \ldots \ldots \ldots (6)$$

where

$$\begin{aligned}a' &= b - c &= a + 1 - \gamma\\ \beta' &= b - d &= \beta + 1 - \gamma\\ \gamma' &= b - a + 1 &= 2 - \gamma\end{aligned} \Bigg\}, \quad \ldots \quad (7)$$

and

$$b = a + 1 - \gamma.$$

Differential Equation of the Hypergeometric Series.

184. If in equation (1) of the preceding article we put $a = 0$, and introduce a, β, and γ in place of b, c, and d by means of equations (4), we obtain

$$\vartheta(\vartheta - 1 + \gamma)y - x(\vartheta + a)(\vartheta + \beta)y = 0, \quad \ldots \quad (1)$$

or, since $\vartheta = x\dfrac{d}{dx}$ and $\vartheta^2 = x^2\dfrac{d^2}{dx^2} + x\dfrac{d}{dx}$, in the ordinary notation

$$x(1-x)\frac{d^2y}{dx^2} + [\gamma - x(1 + a + \beta)]\frac{dy}{dx} - a\beta y = 0, \quad \ldots \quad (2)$$

This is, therefore, the differential equation of the hypergeometric series, $F(a, \beta, \gamma, x)$. Putting, also, $a = 0$ in the value of y_2, we have

$$y = AF(a, \beta, \gamma, x) + Bx^{1-\gamma}F(a + 1 - \gamma, \beta + 1 - \gamma, 2 - \gamma, x)$$

for the complete integral of equation (2).

Since the complete integral of the standard form of the binomial equation of the second order, (1) Art. 183, is the product of this complete integral by x^a, it follows that the general binomial equation of the second order, equation (2), Art. 182, is reducible to the equation of the hypergeometric series in v and z by the transformations $z = qx^s$ and $y = z^a v$.

Integral Values of γ and γ'.

185. When $a = b$ in equation (1), Art. 183, $\gamma = \gamma' = 1$, and the integrals y_1 and y_2 become identical, so that there is but one integral in the form of a hypergeometric series. Again, if a and b differ by an integer, one of the series fails by reason of the occurrence of infinite coefficients. In this case, let a denote the greater of the two quantities, then γ is an integer greater than unity, and γ' is zero or a negative integer.

The coefficient of x^{n-1}, in $F(a', \beta', \gamma', x)$, is

$$\frac{(a+1-\gamma)\ldots(a+n-1-\gamma)(\beta+1-\gamma)\ldots(\beta+n-1-\gamma)}{(n-1)!(2-\gamma)(3-\gamma)\ldots(n-\gamma)}.$$

This is the coefficient of x^{b+n-1}, that is, of $x^{a+n-\gamma}$ in y_2, and is the first which becomes infinite when $\gamma = n$. Now, putting

$$\gamma = n - h,$$

and denoting the sum of the preceding terms of y_2 (which do not become infinite when $h = 0$) by T, the complete integral may be written

$$y = A_0 y_1 + B_0 T + \frac{B}{h} x^{a+h}\left(1 + \frac{(a+h)(\beta+h)}{(1+h)(\gamma+h)}x + \ldots\right), (1)$$

in which $\dfrac{B}{h}$ is the product of B_0 and the coefficient written above, so that, when $h = 0$, B has the finite value

$$B = B_0 \frac{(a+1-n)\ldots(a-1)(\beta+1-n)\ldots(\beta-1)}{(n-1)!(2-n)(3-n)\ldots(-1)}. (2)$$

Putting

$$x^a\left(1 + \frac{(a+h)(\beta+h)}{(1+h)(\gamma+h)}x + \ldots\right) = \psi(h) = x^a \Sigma H_r x^r, (3)$$

§ XVI.]   *INTEGRAL VALUES OF γ AND γ'.*   203

we have, as in Arts. 172 and 175, $y_1 = \psi(0)$; and, expanding in powers of h, equation (1) becomes

$$y = A_0 y_1 + B_0 T + \frac{B}{h}(1 + h \log x + \ldots)[y_1 + h\psi'(0) + \ldots];$$

or, putting A for $A_0 + \frac{B}{h}$ and y' for $\psi'(0)$,

$$y = Ay_1 + B_0 T + By_1 \log x + By' + \ldots, \quad \ldots (4)$$

in which we have retained all the terms which do not vanish with h.

To find y' or $\psi'(0)$, we have, from equation (3),

$$\psi'(h) = x^a \, \Sigma H_r \frac{d \log H_r}{dh} x^r;$$

whence, putting $h = 0$,

$$y' = x^a \left[\frac{a\beta}{1 \cdot \gamma}\left(\frac{1}{a} + \frac{1}{\beta} - \frac{1}{1} - \frac{1}{\gamma}\right) x + \ldots \right] \quad \ldots (5)$$

Finally, writing the complete integral (4) in the form $y = Ay_1 + B\eta$, and taking the value of B_0 from equation (3), we have, for the second integral,

$$\eta = y_1 \log x + (-1)^\gamma \frac{(\gamma - 1)!\,(\gamma - 2)!}{(a+1-\gamma)\ldots(a-1)(\beta+1-\gamma)\ldots(\beta-1)} T + y', \quad (6)$$

where y_1 is the first integral $x^a F(a, \beta, \gamma, x)$, T the terms which do not become infinite in the usual expression for the second integral, and y' the supplementary series given in equation (5).

It is to be noticed that when $\gamma = 1$, $T = 0$.

186. In this general solution of the case in which γ is an integer, the supplementary series y' is the same as the first

integral y_1, except that each coefficient is multiplied by a quantity which may be called its *adjunct*. The adjunct consists of the sum of the reciprocals of the factors in the numerator diminished by the like sum for the factors in the denominator. The first term in y_1 must be regarded as having the adjunct zero.

If y_1 is a finite series, it is to be noticed that the adjunct of each of the vanishing terms is infinite and equal to the reciprocal of the vanishing factor. Thus the corresponding terms of the supplementary series do not vanish, but are precisely as written in the expression for y_1, except that the zero factors in the numerators are omitted.

187. As an illustration, let us take the equation

$$(x^2 - x^3)\frac{d^2y}{dx^2} + (x - x^2)\frac{dy}{dx} - (1 - 9x)y = 0,$$

which, when written in the form (1), Art. 183, is

$$(\vartheta^2 - 1)y - x(\vartheta^2 - 9)y = 0,$$

so that $a = 1, b = -1, c = 3, d = -3$; whence $\alpha = -2, \beta = 4, \gamma = 3$. We have, therefore,

$$y_1 = x\left(1 + \frac{-2\cdot 4}{1\cdot 3}x + \frac{-2\cdot -1\cdot 4\cdot 5}{1\cdot 2\cdot 3\cdot 4}x^2\right)$$
$$+ \frac{-2\cdot -1\cdot 0\cdot 4\cdot 5\cdot 6}{1\cdot 2\cdot 3^2\cdot 4\cdot 5}x^4\left(1 + \frac{1\cdot 7}{4\cdot 6}x + \cdots\right), \quad \cdots \quad (1)$$

in which the terms following the first three vanish. For the other integral, employing equation (6), Art. 185, because γ is an integer, we have

$$\eta = y_1 \log x - \frac{2}{-4\cdot -3\cdot 2\cdot 3}x^{-1}\left(1 + \frac{-4\cdot 2}{-1\cdot 1}x\right) + y',$$

where the next term in the expression for T would be infinite. The part of y' corresponding to the actual terms of y_1 is

$$x\left[\frac{-2.4}{1.3}\left(-\frac{1}{2}+\frac{1}{4}-1-\frac{1}{3}\right)x+\frac{-2.-1.5}{1.2.3}\left(-\frac{1}{2}-1+\frac{1}{5}-1-\frac{1}{2}-\frac{1}{3}\right)x^2\right],$$

and the part corresponding to the vanishing terms in equation (1) is as therein written, with the zero omitted. Thus we have

$$y_1 = x - \frac{8}{3}x^2 + \frac{5}{3}x^3,$$

and

$$\eta = y_1 \log x - \frac{1}{36x} - \frac{2}{9} + y',$$

where

$$y' = \frac{38}{9}x^2 - \frac{47}{9}x^3 + \frac{2}{3}x^4\left[1 + \frac{1.7}{4.6}x + \frac{1.2.7.8}{4.5.6.7}x^2 + \ldots\right].$$

Imaginary Values of a and β.

188. We have assumed the roots a and b of $f(\vartheta) = 0$ to be real, but the roots c and d of $\phi(\vartheta) = 0$ may be imaginary. In that case a and β will be conjugate imaginary quantities, say

$$a = \mu + i\nu, \qquad \beta = \mu - i\nu.$$

The integrals will then take the form

$$y_1 = x^a\left[1 + \frac{\mu^2 + \nu^2}{1.\gamma}x + \frac{(\mu^2 + \nu^2)[(\mu + 1)^2 + \nu^2]}{1.2.\gamma(\gamma + 1)}x^2 + \ldots\right],$$

and

$$y_2 = x^{a+1-\gamma}\left[1 + \frac{(\mu + 1 - \gamma)^2 + \nu^2}{1(2 - \gamma)}x\right.$$
$$\left. + \frac{[(\mu + 1 - \gamma)^2 + \nu^2][(\mu + 2 - \gamma)^2 + \nu^2]}{1.2(2 - \gamma)(3 - \gamma)}x^2 + \ldots\right].$$

·Again, when γ is an integer, making the same substitutions in equation (6), Art. 185, the second integral becomes

$$\eta = y_1 \log x + (-1)^\gamma \frac{(\gamma-1)!(\gamma-2)!}{[(\mu+1-\gamma)^2+\nu^2]\ldots[(\mu-1)^2+\nu^2]} T + y',$$

where

$$y' = x^a \left[\frac{\mu^2+\nu^2}{1\cdot\gamma}\left(\frac{2\mu}{\mu^2+\nu^2} - 1 - \frac{1}{\gamma}\right)x + \ldots\right].$$

Infinite Values of a and β.

189. As explained in Art. 165, the function $f(\vartheta)$ must be of the second degree, but $\phi(\vartheta)$ may be of the first degree, the equation being of the form

$$(\vartheta - a)(\vartheta - b)y - x(\vartheta - c)y = 0. \quad \ldots \quad (1)$$

The solution of this equation is included in the general solution already given, for the equation is the result of making d infinite in

$$(\vartheta - a)(\vartheta - b)y - \frac{x}{a-d}(\vartheta - c)(\vartheta - d)y = 0. \quad . \quad (2)$$

Here $\dfrac{x}{a-d}$, that is, $\dfrac{x}{\beta}$ takes the place of x in the standard form; hence equation (5) Art. 183, gives the integral

$$\frac{x^a}{\beta^a}\left(1 + \frac{a\beta}{1\cdot\gamma}\frac{x}{\beta} + \frac{a(a+1)\beta(\beta+1)}{1\cdot 2\cdot\gamma(\gamma+1)}\frac{x^2}{\beta^2} + \ldots\right)$$

for equation (2). Multiplying by the constant β^a, and then making β infinite, we have for the first integral of equation (1)

$$y_1 = x^a\left(1 + \frac{a}{1\cdot\gamma}x + \frac{a(a+1)}{1\cdot 2\cdot\gamma(\gamma+1)}x^2 + \ldots\right)$$

$$= x^a F\left(a, \beta, \gamma, \frac{x}{\beta}\right), \qquad \beta = \infty.$$

In like manner, for the second integral, we obtain

$$y_2 = x^{a+1-\gamma} F\left(a + 1 - \gamma, \beta, 2 - \gamma, \frac{x}{\beta}\right), \quad \beta = \infty.$$

190. Again, when $\phi(\vartheta)$ is a mere constant, the equation being reducible to the form

$$(\vartheta - a)(\vartheta - b)y - xy = 0, \quad \ldots \ldots (1)$$

it is the result of making both c and d infinite in

$$(\vartheta - a)(\vartheta - b)y - \frac{x}{(a-c)(a-d)}(\vartheta - c)(\vartheta - d)y = 0. \quad (2)$$

We have now for the first integral of equation (2)

$$\frac{x^a}{a^a \beta^a}\left(1 + \frac{a\beta}{1.\gamma}\frac{x}{a\beta} + \frac{a(a+1)\beta(\beta+1)}{1.2.\gamma(\gamma+1)}\frac{x^2}{a^2\beta^2} + \cdots\right).$$

Multiplying by $a^a \beta^a$, and putting $a = \infty$, $\beta = \infty$, the first integral of equation (1) is

$$y_1 = x^a\left(1 + \frac{1}{1.\gamma}x + \frac{1}{1.2.\gamma(\gamma+1)}x^2 + \cdots\right)$$

$$= x^a F\left(a, \beta, \gamma, \frac{x}{a\beta}\right), \quad a = \infty, \quad \beta = \infty,$$

and, in like manner, the second integral is

$$y_2 = x^{a+1-\gamma} F\left(a, \beta, 2 - \gamma, \frac{x}{a\beta}\right), \quad a = \infty, \beta = \infty.$$

If, in either of these cases, γ is an integer, so that the logarithmic form of solution is required, the second integral is given by equation (6), Art. 185, and is of the same form, except that the infinite factors disappear after multiplication by β^a or $a^a \beta^a$, and the reciprocals of these factors vanish from the adjuncts in the supplementary series (5).

Cases in which α or β equals γ or Unity.

191. The binomial equation of the first order may be reduced to the form

$$(\vartheta - a)y - x(\vartheta - c)y = 0, \quad \ldots \quad (1)$$

and, with the notation of the preceding articles, its solution in series is

$$y = x^a\left(1 + \frac{a}{1}x + \frac{a(a+1)}{1.2}x^2 + \ldots\right) \ldots \quad (2)$$

This is, of course, the value of $x^a(1-x)^{c-a}$, or $x^a(1-x)^{-a}$, which is the integral in its ordinary finite form. The series involved may obviously be written $F(a, \gamma, \gamma, x)$, where the value of γ is arbitrary, and accordingly this value of y is one integral of the equation

$$(\vartheta - a)(\vartheta - b)y - x(\vartheta - c)(\vartheta - b + 1)y = 0, \quad \ldots \quad (3)$$

since $\beta = \gamma$ in equations (4), Art. 183, makes $d = b - 1$. The other integral of this equation is

$$y_2 = x^b\left(1 + \frac{b-c}{b-a+1}x + \frac{(b-c)(b-c+1)}{(b-a+1)(b-a+2)}x^2 + \ldots\right), \quad (4)$$

or

$$y_2 = x^{a+1-\gamma}F(a + 1 - \gamma, 1, 2 - \gamma, x).$$

192. Equation (3) might have been solved by the method of Art. 141; for it becomes an exact differential equation when multiplied by x^{-b-1} [see equation (1), Art. 140]. The result of the first integration is

$$(\vartheta - a)y - x(\vartheta - c)y = Cx^b;$$

and in the second integration the value of y in equation (2) is the complementary function, and that of y_2 is the particular integral. Thus the hypergeometric series in which one of the first two elements is equal to γ reduces to the form assumed when the equation is of the first order, and that in which one of the first two elements is unity is of the form of the particular integral of an equation of the first order when the second member is a power of x.

The Binomial Equation of the Third Order.

193. The binomial equation of the third order may be reduced to the form

$$(\vartheta - a)(\vartheta - b)(\vartheta - c)y - x(\vartheta - d)(\vartheta - e)(\vartheta - f)y = 0.$$

One of its three independent integrals is

$$y_1 = x^a\left(1 + \frac{\alpha\beta\gamma}{1.\delta\epsilon}x + \frac{\alpha(\alpha+1)\beta(\beta+1)\gamma(\gamma+1)}{1.2.\delta(\delta+1)\epsilon(\epsilon+1)}x^2 + \ldots\right),$$

where
$$\alpha = a - d, \quad \beta = a - e, \quad \gamma = a - f,$$
$$\delta = a - b + 1, \quad \epsilon = a - c + 1,$$

and the other two are the result of interchanging a and b, and a and c respectively.*

The notation $F\left(\genfrac{}{}{0pt}{}{\alpha, \beta, \gamma,}{\delta, \epsilon,} x,\right)$ has been employed for the series involved in the value of y_1 above.

---

* When two of the roots a, b, and c of $f(\vartheta)$ differ by an integer, so that one of the quantities δ or ϵ is an integer, the powers of x which occur in one of the three integrals will occur in another with infinite coefficients. By the process employed in Art. 185 these infinite terms are replaced by terms involving $\log x$ and the adjuncts. If both δ and ϵ are integers, the third integral contains terms which occur in each of the others, with doubly infinite coefficients, and by a similar process these may be replaced by terms involving $(\log x)^2$ as well as $\log x$. Similar results hold for binomial equations of any order. See *American Journal of Mathematics*, vol. xi., pp. 49, 50, 51.

Development of the Solution in Descending Series.

194. When both of the functions f and ϕ in the binomial equation are of the second degree, that is, when α and β are finite, the integrals y_1 and y_2 are convergent for values of x less than unity, and divergent when x is greater than unity. In the latter case, convergent series are obtained by developing in descending powers of x, or what is the same thing, ascending powers of x^{-1}. Putting, in equation (1), Art. 183,

$$z = \frac{1}{x}, \quad \text{whence} \quad \vartheta' = z\frac{d}{dz} = -\vartheta,$$

we have

$$(\vartheta' + c)(\vartheta' + d)y - z(\vartheta' + a)(\vartheta' + b)y = 0;$$

hence the results are obtained by changing a, b, c, and d, in the preceding results, to $-c$, $-d$, $-a$, and $-b$. Making these changes in equations (4), and denoting the new values of α, β, and γ by α_1, β_1, and γ_1, we find

$$\alpha_1 = -c + a \quad = \alpha,$$
$$\beta_1 = -c + b \quad = \alpha + 1 - \gamma,$$
$$\gamma_1 = -c + d + 1 = \alpha + 1 - \beta;$$

and the integrals are

$$Y_1 = z^{-c}F(\alpha_1, \beta_1, \gamma_1, z)$$
$$= x^c F\left(\alpha,\ \alpha + 1 - \gamma,\ \alpha + 1 - \beta,\ \frac{1}{x}\right), \quad \ldots \quad (1)$$
$$Y_2 = z^{-d}F(\alpha_1', \beta_1', \gamma_1', z)$$
$$= x^d F\left(\beta,\ \beta + 1 - \gamma,\ \beta + 1 - \alpha,\ \frac{1}{x}\right). \quad \ldots \quad (2)$$

Transformation of the Equation of the Hypergeometric Series.

195. The equation of the hypergeometric series,

$$x(1-x)\frac{d^2y}{dx^2} + [\gamma - x(1 + \alpha + \beta)]\frac{dy}{dx} - \alpha\beta y = 0, \quad . \quad (1)$$

admits of transformation in a variety of ways into equations of the same form, leading to other integrals still expressed by means of hypergeometric series. One such transformation is obviously $y = x^{1-\gamma}v$; for, since

$$y_2 = x^{1-\gamma}F(\alpha', \beta', \gamma', x),$$

this will give an equation for v one of whose integrals is the simple hypergeometric series $F(\alpha', \beta', \gamma', x)$, that is, the transformed equation is of the form (1), the new values of α, β, and γ being

$$\alpha' = \alpha + 1 - \gamma,$$
$$\beta' = \beta + 1 - \gamma,$$
$$\gamma' = 2 - \gamma.$$

The second integral of this equation will be

$$v_2 = x^{1-\gamma'}F(\alpha' + 1 - \gamma', \beta' + 1 - \gamma', 2 - \gamma', x)$$
$$= x^{\gamma-1}F(\alpha, \beta, \gamma, x),$$

and the corresponding value of y is $F(\alpha, \beta, \gamma, x)$, which is the value of y_1. This transformation, therefore, gives no new integral.

196. Let us now make a transformation of the form

$$y = (1-x)^\mu v.$$

Comparing equation (1) with the form

$$\frac{d^2y}{dx^2} + P\frac{dy}{dx} + Qy = 0,$$

we have

$$P = \frac{\gamma - x(1 + \alpha + \beta)}{x(1 - x)} = \frac{\gamma}{x} + \frac{\gamma - \alpha - \beta - 1}{1 - x},$$

and

$$Q = -\frac{\alpha\beta}{x(1 - x)}.$$

Hence, putting in the formulæ of Art. 152 $w_1 = (1 - x)^\mu$, we find, for the transformed equation,

$$P_1 = \frac{\gamma}{x} + \frac{\gamma - \alpha - \beta - 1 - 2\mu}{1 - x},$$

$$Q_1 = \frac{\mu(\mu - 1)}{(1 - x)^2} - \frac{\mu}{1 - x}P + Q$$

$$= \frac{\mu(\mu - 1) - \mu(\gamma - \alpha - \beta - 1)}{(1 - x)^2} - \frac{\mu\gamma + \alpha\beta}{x(1 - x)}.$$

In order that Q_1 may take the same form as Q, let μ be so determined that the first term of this expression vanishes. This gives $\mu = 0$ (in which case no transformation is effected), or else

$$\mu = \gamma - \alpha - \beta.$$

Then, if $\alpha_1, \beta_1, \gamma_1$ have the same relation to P_1 and Q_1 that α, β, γ have to P and Q, the form of P_1 shows that $\gamma_1 = \gamma$, and

$$\alpha_1 + \beta_1 = \alpha + \beta + 2\mu,$$

and that of Q_1 shows that

$$\alpha_1\beta_1 = \alpha\beta + \mu\gamma.$$

Substituting the value of μ above, and solving, we find

$$a_1 = \gamma - a,$$
$$\beta_1 = \gamma - \beta,$$
$$\gamma_1 = \gamma.$$

The integrals of the transformed equation are, therefore,

$$v_1 = F(\gamma - a, \gamma - \beta, \gamma, x),$$

and

$$v_2 = x^{1-\gamma} F(1 - a, 1 - \beta, 2 - \gamma, x),$$

v_2 being derived from v_1 by the same changes which convert y_1 into y_2. Denoting the corresponding values of y by y_3 and y_4, we have thus four integrals involving hypergeometric series of which the variable element is x; namely,

$$y_1 = F(a, \beta, \gamma, x),$$
$$y_2 = x^{1-\gamma} F(a + 1 - \gamma, \beta + 1 - \gamma, 2 - \gamma, x),$$
$$y_3 = (1 - x)^{\gamma-a-\beta} F(\gamma - a, \gamma - \beta, \gamma, x),$$
$$y_4 = x^{1-\gamma} (1 - x)^{\gamma-a-\beta} F(1 - a, 1 - \beta, 2 - \gamma, x).$$

197. The integral y_3 is the product of two series involving powers of x with positive integral exponents, each of which has unity for its first term, and is convergent when $x < 1$. It follows that y_3 is a series of the same form. But, from the process of integration in series, we know that there can be but one integral of this form, namely y_1. Hence we have the theorem, $y_1 = y_3$, or

$$F(a, \beta, \gamma, x) = (1 - x)^{\gamma-a-\beta} F(\gamma - a, \gamma - \beta, \gamma, x).$$

In like manner $y_2 = y_4$.

Change of the Independent Variable.

198. It is obvious from the form of the equation of the hypergeometric series, equation (1), Art. 195, that, if we change the independent variable from x to

$$t = 1 - x,$$

the first and third terms will be unchanged, and the second will be unchanged in form. The result is

$$t(1-t)\frac{d^2y}{dt^2} + [1+\alpha+\beta-\gamma - t(1+\alpha+\beta)]\frac{dy}{dt} - \alpha\beta y = 0;$$

comparing this with the original equation, we find that α and β are unchanged, while γ is replaced by

$$1 + \alpha + \beta - \gamma.$$

We hence derive the integral

$$y_5 = F(\alpha, \beta, 1+\alpha+\beta-\gamma, 1-x).$$

A comparison of this integral with y_1 or $F(\alpha, \beta, \gamma, x)$ shows that from any integral expressed by a hypergeometric series we can derive another integral of the same equation by making the above-mentioned change in the third element, and at the same time changing the fourth or variable element to what may be called its *complement*, that is, the result of subtracting it from unity. The process applies equally well to an integral of the form $y = wv$, where v is a hypergeometric series; for a new integral of the equation for v gives a new integral of the equation for y. Thus the integrals y_2, y_3, and y_4 would lead to the three new integrals y_7, y_6, and y_8, which will be found in a subsequent article.

§ XVI.] CHANGE OF THE INDEPENDENT VARIABLE. 215

199. It is shown in Art. 194 that, when we change the independent variable to $z = \dfrac{1}{x}$, the binomial equation retains its form. The equation considered in that article becomes that of the hypergeometric series when we put $a = 0$, whence $c = -a$. Thus equation (1), Art. 194, gives the integral

$$Y_1 = x^{-a} F\left(a,\ a+1-\gamma,\ a+1-\beta,\ \dfrac{1}{x}\right).$$

A comparison of this integral with $F(a, \beta, \gamma, x)$ gives a method by which we may pass from any integral in the form of a hypergeometric series to another in which the variable element is replaced by its reciprocal; and, as in the preceding article, the process applies also to an integral of the form $y = wv$, where v is a hypergeometric series.

200. If we start with the variable x, and alternately take the complement and the reciprocal, we obtain the following six values of the variable,

$$x,\ 1-x,\ \dfrac{1}{1-x},\ -\dfrac{x}{1-x},\ -\dfrac{1-x}{x},\ \dfrac{1}{x},$$

the seventh term of the series being identical with the first. The corresponding integrals derived from y, by the processes of Arts. 198 and 199, are

$$y_1 = F(a,\ \beta,\ \gamma,\ x),$$

$$y_5 = F(a,\ \beta,\ 1+a+\beta-\gamma,\ 1-x),$$

$$y_9 = (1-x)^{-a} F\left(a,\ \gamma-\beta,\ a+1-\beta,\ \dfrac{1}{1-x}\right),$$

$$y_{13} = (1-x)^{-a} F\left(a,\ \gamma-\beta,\ \gamma,\ -\dfrac{x}{1-x}\right),$$

$$y_{17} = x^{-a} F\left(a,\ a+1-\gamma,\ 1+a+\beta-\gamma,\ -\dfrac{1-x}{x}\right),$$

$$y_{21} = x^{-a} F\left(a,\ a+1-\gamma,\ a+1-\beta,\ \dfrac{1}{x}\right),$$

where, in writing the last two, the constant factor $(-1)^{-a}$ has been omitted.

201. From each of the integrals given above we may derive three others, exactly as y_2, y_3 and y_4 are derived from y_1. We have thus the following system of twenty-four integrals,

$y_1 = F(a, \beta, \gamma, x)$,

$y_2 = x^{1-\gamma} F(a+1-\gamma, \beta+1-\gamma, 2-\gamma, x)$,

$y_3 = (1-x)^{\gamma-a-\beta} F(\gamma-a, \gamma-\beta, \gamma, x)$,

$y_4 = x^{1-\gamma}(1-x)^{\gamma-a-\beta} F(1-a, 1-\beta, 2-\gamma, x)$,

$y_5 = F(a, \beta, 1+a+\beta-\gamma, 1-x)$,

$y_6 = (1-x)^{\gamma-a-\beta} F(\gamma-a, \gamma-\beta, 1-a-\beta+\gamma, 1-x)$;

$y_7 = x^{1-\gamma} F(a+1-\gamma, \beta+1-\gamma, 1+a+\beta-\gamma, 1-x)$,

$y_8 = x^{1-\gamma}(1-x)^{\gamma-a-\beta} F(1-a, 1-\beta, 1-a-\beta+\gamma, 1-x)$,

$y_9 = (1-x)^{-a} F\left(a, \gamma-\beta, a+1-\beta, \dfrac{1}{1-x}\right)$,

$y_{10} = (1-x)^{-\beta} F\left(\beta, \gamma-a, \beta+1-a, \dfrac{1}{1-x}\right)$,

$y_{11} = (-x)^{1-\gamma}(1-x)^{\gamma-1-a} F\left(1-\beta, a+1-\gamma, a+1-\beta, \dfrac{1}{1-x}\right)$,

$y_{12} = (-x)^{1-\gamma}(1-x)^{\gamma-1-\beta} F\left(1-a, \beta+1-\gamma, \beta+1-a, \dfrac{1}{1-x}\right)$,

$y_{13} = (1-x)^{-a} F\left(a, \gamma-\beta, \gamma, -\dfrac{x}{1-x}\right)$,

$y_{14} = x^{1-\gamma}(1-x)^{\gamma-1-a} F\left(a+1-\gamma, 1-\beta, 2-\gamma, -\dfrac{x}{1-x}\right)$,

$y_{15} = (1-x)^{-\beta} F\left(\gamma-a, \beta, \gamma, -\dfrac{x}{1-x}\right)$,

$y_{16} = x^{1-\gamma}(1-x)^{\gamma-1-\beta} F\left(1-a, \beta+1-\gamma, 2-\gamma, -\dfrac{x}{1-x}\right)$,

$$y_{17} = x^{-a}F\left(a,\ a+1-\gamma,\ 1+a+\beta-\gamma,\ -\frac{1-x}{x}\right),$$

$$y_{18} = x^{\beta-\gamma}(1-x)^{\gamma-a-\beta}F\left(\gamma-\beta,\ 1-\beta,\ 1-a-\beta+\gamma,\ -\frac{1-x}{x}\right),$$

$$y_{19} = x^{-\beta}F\left(\beta+1-\gamma,\ \beta,\ 1+a+\beta-\gamma,\ -\frac{1-x}{x}\right),$$

$$y_{20} = x^{a-\gamma}(1-x)^{\gamma-a-\beta}F\left(1-a,\ \gamma-a,\ 1-a-\beta+\gamma,\ -\frac{1-x}{x}\right),$$

$$y_{21} = x^{-a}F\left(a,\ a+1-\gamma,\ a+1-\beta,\ \frac{1}{x}\right),$$

$$y_{22} = x^{-\beta}F\left(\beta,\ \beta+1-\gamma,\ \beta+1-a,\ \frac{1}{x}\right),$$

$$y_{23} = x^{\beta-\gamma}(x-1)^{\gamma-a-\beta}F\left(1-\beta,\ \gamma-\beta,\ a+1-\beta,\ \frac{1}{x}\right),$$

$$y_{24} = x^{a-\gamma}(x-1)^{\gamma-a-\beta}F\left(1-a,\ \gamma-a,\ \beta+1-a,\ \frac{1}{x}\right).$$

Since the binomial equation of the second order can be transformed into the equation of the hypergeometric series, it follows that the binomial equation has in general twenty-four integrals expressible by means of hypergeometric series.* But, in the cases considered in Arts. 189 and 190, where a or β is infinite, we have only the integrals y_1 and y_2.

---

* The twenty-four integrals are written above exactly as they arise in the process indicated, except that the factor $(-1)^{1-\gamma}$ is dropped in the case of y_{14} and y_{16}, and $(-1)^{\gamma-a-\beta}$ is dropped in the case of y_{18} and y_{20}. Because $y_1 = y_3$ and $y_2 = y_4$, the first and third integral of each group are equal, and so also are the second and fourth, the omission of a factor in the cases mentioned above causing no exception. It may also be shown, by comparing the developments in powers of x, that the integrals of the first group are respectively equal to those of the fourth group, and those of the second to those of the fifth group. But in the third and sixth groups $y_9 = (-1)^a y_{21}$ and $y_{10} = (-1)^\beta y_{22}$. Thus the twenty-four integrals consist of six sets of equal quantities, as follows:—

Solutions in Finite Form.

202. The condition that $F(\alpha, \beta, \gamma, x)$ may represent a finite series is readily seen to be that one of the elements α or β shall be zero or a negative integer. But, since $y_1 = y_3$, the form of y_3 shows that, if either $\gamma - \alpha$ or $\gamma - \beta$ is zero* or a negative integer, $F(\alpha, \beta, \gamma, x)$ may be expressed in finite algebraic form. For example, one integral of the equation

$$2x(1-x)\frac{d^2y}{dx^2} + [1-(2n+5)x]\frac{dy}{dx} - 3ny = 0$$

is the infinite series represented by $F(\frac{3}{2}, n, \frac{1}{2}, x)$. Here $\gamma - \alpha$ is a negative integer, and, using the form y_3, the integral may be written

$$(1-x)^{-n-1}F(-1, \tfrac{1}{2}-n, \tfrac{1}{2}, x),$$

$$y_1 = y_3 = y_{13} = y_{15},$$
$$y_2 = y_4 = y_{14} = y_{16},$$
$$y_5 = y_7 = y_{17} = y_{19},$$
$$y_6 = y_8 = y_{18} = y_{20},$$
$$y_9 = y_{11} = (-1)^\alpha y_{21} = (-1)^\alpha y_{23},$$
$$y_{10} = y_{12} = (-1)^\beta y_{22} = (-1)^\beta y_{24}.$$

Between any three integrals belonging to different sets there must exist a relation of the form $y_p = My_r + Ny_s$. These relations, in which the values of M and N involve Gamma Functions, are equivalent to those given by Gauss in the memoir "Determinatio Seriei Nostrae per Aequationem Differentialem Secundi Ordinis," *Werke*, vol. iii. See equations [86], p. 213, and [93], p. 220. The twenty-four integrals, and their separation into six sets of equal quantities, were first given by Kummer, in a memoir "Ueber die hypergeometrische Reihe," *Crelle*, vol. xv., p. 52. The order of the integrals is different from that given above, and some errors involving factors of the form $(-1)^\mu$ occur in the statement of the equalities. The values of M and N are given by Kummer for the integrals numbered by him 1, 3, 5, 7, 13, and 14, corresponding to the integrals y_1, y_2, y_5, y_6, y_9, and y_{10} above.

* The case in which $\gamma - \beta = 0$ has already been considered in Art. 191.

in which the second factor is the finite series

$$1 + \frac{-1(\frac{1}{2} - n)}{1 \cdot \frac{1}{2}} x = 1 + (2n - 1)x.$$

Hence the integral in question is

$$F(\tfrac{3}{2}, n, \tfrac{1}{2}, x) = \frac{1 + (2n - 1)x}{(1 - x)^{n+1}}.$$

203. In like manner the integral y_2 will be a finite series if either of the quantities $\alpha + 1 - \gamma$ or $\beta + 1 - \gamma$ is zero or a negative integer; and, since $y_2 = y_4$, the form of y_4 shows that if either $1 - \alpha$ or $1 - \beta$ is zero or a negative integer (in other words, if α or β is a positive integer), y_2 may be expressed in finite form. It will be noticed that the eight quantities,

$$\alpha,\ \beta,\ \gamma - \alpha,\ \gamma - \beta,\ \alpha + 1 - \gamma,\ \beta + 1 - \gamma,\ 1 - \alpha,\ 1 - \beta,$$

are the only values of the first two elements in the twenty-four integrals; hence the only cases in which they furnish finite integrals are those in which either α, β, $\gamma - \alpha$, or $\gamma - \beta$ is an integer.

In the case of the general binomial equation of the second order, the condition given in the preceding article, when applied to both integrals, is sufficient to determine whether finite algebraic solutions exist.*

---

* Finite solutions involving transcendental functions occur in certain cases considered in the following chapter. See Arts. 209, 213, 214, and 217.

Examples XVI.

1. Show that, in the notation of the hypergeometric series,

$$(t + u)^n + (t - u)^n = 2t^n F\left(-\tfrac{1}{2}n, -\tfrac{1}{2}n + \tfrac{1}{2}, \tfrac{1}{2}, \frac{u^2}{t^2}\right),$$

$$(t + u)^n - (t - u)^n = 2nt^{n-1}u F\left(-\tfrac{1}{2}n + \tfrac{1}{2}, -\tfrac{1}{2}n + 1, \tfrac{3}{2}, \frac{u^2}{t^2}\right),$$

$$\log(1 + x) = x F(1, 1, 2, -x),$$

$$\log\frac{1+x}{1-x} = 2x F(\tfrac{1}{2}, 1, \tfrac{3}{2}, x^2),$$

$$e^x = F\left(1, k, 1, \frac{x}{k}\right) = 1 + x F\left(1, k, 2, \frac{x}{k}\right)$$
$$= 1 + x + \tfrac{1}{2}x^2 F\left(1, k, 3, \frac{x}{k}\right) = \text{etc., where } k = \infty,$$

$$\sin x = x F\left(k, k', \tfrac{3}{2}, -\frac{x^2}{4kk'}\right), \qquad k = k' = \infty,$$

$$\cos x = F\left(k, k', \tfrac{1}{2}, -\frac{x^2}{4kk'}\right), \qquad k = k' = \infty,$$

$$\cosh x = F\left(k, k', \tfrac{1}{2}, \frac{x^2}{4kk'}\right), \qquad k = k' = \infty,$$

$$\sin^{-1} x = x F(\tfrac{1}{2}, \tfrac{1}{2}, \tfrac{3}{2}, x^2),$$

$$\tan^{-1} x = x F(\tfrac{1}{2}, 1, \tfrac{3}{2}, -x^2).$$

2. Show that

$$\frac{d}{dx} F(\alpha, \beta, \gamma, x) = \frac{\alpha\beta}{\gamma} F(\alpha + 1, \beta + 1, \gamma + 1, x),$$

$$\frac{d^2}{dx^2} F(\alpha, \beta, \gamma, x) = \frac{\alpha(\alpha + 1)\beta(\beta + 1)}{\gamma(\gamma + 1)} F(\alpha + 2, \beta + 2, \gamma + 2, x), \text{etc.}$$

3. Show that the equation

$$Ay + (B + Cx)\frac{dy}{dx} + (D + Ex + Fx^2)\frac{d^2y}{dx^2} = 0$$

can be reduced to the equation of the hypergeometric series, and hence that the complete integral is

$$y = c_1 F\left(\alpha, \beta, \gamma, \frac{x-a}{b-a}\right) + c_2 F\left(\alpha, \beta, \gamma', \frac{x-b}{a-b}\right),$$

where a and b are the roots of $D + Ex + Fx^2 = 0$, $\alpha\beta = \dfrac{A}{F}$,

$\alpha + \beta + 1 = \dfrac{C}{F}$, $\gamma = \dfrac{B + aC}{(a-b)F}$ and $\gamma' = \dfrac{B + bC}{(b-a)F}$, the two independent integrals being related as y_1 is to y_5 in Art. 198.

4. Find the particular integral of the equation

$$(\vartheta - a)(\vartheta - b)y - x(\vartheta - c)(\vartheta - d)y = kx^p,$$

and derive the integrals in Art. 183 from the result.

$$Y = \frac{kx^p}{(p-a)(p-b)}\left[1 + \frac{(p-c)(p-d)}{(p-a+1)(p-b+1)}x + \cdots\right].$$

Solve the following equations:—

5. $x(1-x)\dfrac{d^2y}{dx^2} + (\tfrac{3}{2} - 2x)\dfrac{dy}{dx} - \tfrac{1}{4}y = 0$,

$$y = AF(\tfrac{1}{2}, \tfrac{1}{2}, \tfrac{3}{2}, x) + Bx^{-\tfrac{1}{2}} = \frac{A\sin^{-1}\sqrt{x} + B}{\sqrt{x}}.$$

6. $2x(1-x)\dfrac{d^2y}{dx^2} + x\dfrac{dy}{dx} - y = 0$,

$$y = x(A + B\log x) + B\left(2 + \frac{1}{4}x^2 + \frac{1.3}{4.6}\frac{x^3}{2} + \frac{1.3.5}{4.6.8}\frac{x^4}{3} + \cdots\right).$$

7. Transform the series

$$y = 1 + 2\frac{8}{9}x + 3\frac{8.10}{9.11}x^2 + 4\frac{8.10.12}{9.11.13}x^3 + \cdots$$

by means of the theorem of Art. 197.

$$y = (1-x)^{-\frac{3}{2}}\left(1 + \frac{1.5}{2.9}x + \frac{1.3.5.7}{2.4.9.11}x^2 + \cdots\right).$$

Solve, in finite form, the following equations: —

8. $2x(1-x)\frac{d^2y}{dx^2} + (1-11x)\frac{dy}{dx} - 10y = 0$,

$$y = A\frac{1 + 6x + x^2}{(1-x)^4} + B\frac{\sqrt{x}(1+x)}{(1-x)^4}.$$

9. $x(1-x)\frac{d^2y}{dx^2} + \frac{1}{3}(1-2x)\frac{dy}{dx} + \frac{20}{9}y = 0$,

$$y = A(1-x)^{\frac{2}{3}}(1-6x) + Bx^{\frac{2}{3}}(5-6x).$$

10. $2x(1-x)\frac{d^2y}{dx^2} + \frac{dy}{dx} + 4y = 0$,

$$y = A(3 - 12x + 8x^2) + Bx^{\frac{1}{2}}(1-x)^{\frac{3}{2}}.$$

11. Solve the equation

$$4\frac{d^2y}{dx^2} + 3\frac{2-x^2}{(1-x^2)^2}y = 0,$$

first transforming to the new independent variable $z = 1 - x^2$.

$$y = A(1-x^2)^{\frac{3}{4}} + Bx(1-x^2)^{\frac{1}{4}}.$$

12. When a is a negative integer, the six integrals of Art. 200 are all finite series, and therefore must, in that case, be all multiples of the integral y_1. Verify this when $a = -1$.

13. Show, by comparing the first two terms of the development, that $y_1 = y_{13}$, and thence that

$$F(a, \beta, \gamma, \sin^2\theta) = (\cos^2\theta)^{\gamma-a-\beta}F(\gamma-a, \gamma-\beta, \gamma, \sin^2\theta)$$
$$= (\sec^2\theta)^a F(a, \gamma-\beta, \gamma, -\tan^2\theta)$$
$$= (\sec^2\theta)^\beta F(\gamma-a, \beta, \gamma, -\tan^2\theta).$$

§ XVI.] EXAMPLES. 223

14. From the expression for $\sin^{-1} x$ as a hypergeometric series, derive

$$\theta = \sin\theta F(\tfrac{1}{2}, \tfrac{1}{2}, \tfrac{3}{2}, \sin^2\theta)$$
$$= \sin\theta\cos\theta F(1, 1, \tfrac{3}{2}, \sin^2\theta)$$
$$= \tan\theta F(\tfrac{1}{2}, 1, \tfrac{3}{2}, -\tan^2\theta).$$

15. The integrals of the equation

$$\frac{d^2y}{d\theta^2} + n^2 y = 0$$

are $\sin n\theta$ and $\cos n\theta$; form the equation in which $x = \sin\theta$ is the independent variable, and thence derive four expressions, as in Ex. 13, for each of these quantities.

$$\sin n\theta = n\sin\theta F(\tfrac{1}{2} - \tfrac{1}{2}n, \tfrac{1}{2} + \tfrac{1}{2}n, \tfrac{3}{2}, \sin^2\theta),$$
$$= n\sin\theta\cos\theta F(1 + \tfrac{1}{2}n, 1 - \tfrac{1}{2}n, \tfrac{3}{2}, \sin^2\theta),$$
$$= n\sin\theta(\cos\theta)^{n-1} F(1 - \tfrac{1}{2}n, \tfrac{1}{2} - \tfrac{1}{2}n, \tfrac{3}{2}, -\tan^2\theta),$$
$$= n\sin\theta(\cos\theta)^{-n-1} F(1 + \tfrac{1}{2}n, \tfrac{1}{2} + \tfrac{1}{2}n, \tfrac{3}{2}, -\tan^2\theta);$$
$$\cos n\theta = F(-\tfrac{1}{2}n, \tfrac{1}{2}n, \tfrac{1}{2}, \sin^2\theta),$$
$$= \cos\theta F(\tfrac{1}{2} + \tfrac{1}{2}n, \tfrac{1}{2} - \tfrac{1}{2}n, \tfrac{1}{2}, \sin^2\theta),$$
$$= (\cos\theta)^n F(-\tfrac{1}{2}n, \tfrac{1}{2} - \tfrac{1}{2}n, \tfrac{1}{2}, -\tan^2\theta),$$
$$= (\cos\theta)^{-n} F(\tfrac{1}{2}n, \tfrac{1}{2} + \tfrac{1}{2}n, \tfrac{1}{2}, -\tan^2\theta).$$

16. Denoting by R the expression

$$x(x^4 - 1)\frac{d^2t}{dx^2} + (3x^4 - 1)\frac{dt}{dx} + x^3 t,$$

show that the equation $xt\dfrac{dR}{dx} + \left(t + 3x\dfrac{dt}{dx}\right)R = 0$ is equivalent to

$$x^2(x^4 - 1)\frac{d^3u}{dx^3} + 3x(3x^4 - 1)\frac{d^2u}{dx^2} + (19x^4 - 1)\frac{du}{dx} + 8x^3 u = 0,$$

where $u = t^2$; and thence that

$$1 + \frac{1^3}{2^3}z + \frac{1^3 \cdot 3^3}{2^3 \cdot 4^3}z^2 + \ldots = \left(1 + \frac{1^2}{4^2}z + \frac{1^2 \cdot 5^2}{4^2 \cdot 8^2}z^2 + \ldots\right)^2.$$

—Gauss, *Werke*, vol. iii. p. 424.

CHAPTER IX.

SPECIAL FORMS OF DIFFERENTIAL EQUATIONS.

XVII.

Riccati's Equation.

204. There are certain forms of differential equations which, either for their historic interest or their importance in mathematical physics, deserve special consideration. Of these we shall consider first Riccati's equation and its transformations.

The equation

$$\frac{dy}{dx} + by^2 = cx^m \quad \cdots \cdots \cdots (1)$$

was first discussed by Riccati, and attracted attention from the fact that it was shown to be integrable in a finite form for certain values of m. If we put $\frac{x}{b}$ in place of x, and write a^2 for the constant $\frac{c}{b^{m+1}}$, the equation becomes

$$\frac{dy}{dx} + y^2 = a^2 x^m, \quad \cdots \cdots \cdots (2)$$

so that no generality is lost by assuming the coefficient b equal to unity. The case in which the coefficient of x^m is negative will be provided for by changing a^2 to $-a^2$, that is, a to ia, in the results.

205. In the form (2), Riccati's equation is the equation of the first order connected, as in Art. 151, with the linear equation of the second order,

$$\frac{d^2u}{dx^2} - a^2 x^m u = 0; \quad \ldots \ldots \ldots (3)$$

in other words, this last equation is the result of the substitution

$$y = \frac{1}{u}\frac{du}{dx}$$

in equation (2); and, denoting its complete integral by

$$u = c_1 X_1 + c_2 X_2, \quad \ldots \ldots \ldots (4)$$

that of equation (2) is

$$y = \frac{c_1 X_1' + c_2 X_2'}{c_1 X_1 + c_2 X_2} = \frac{X_1' + c X_2'}{X_1 + c X_2}, \quad \ldots \ldots (5)$$

which shows the manner in which the constant of integration enters the solution.

Standard Linear Form of the Equation.

206. The discussion of Riccati's equation is simplified by using the linear form (3); moreover, the expression of the results and transformation to other important forms is facilitated by writing the exponent m in the form $2q - 2$.* We shall, therefore, take

$$\frac{d^2u}{dx^2} - a^2 x^{2q-2} u = 0 \quad \ldots \ldots \ldots (1)$$

as the standard form of Riccati's equation from which to deter-

---

* This improvement of the notation was introduced by Cayley, *Philosophical Magazine*, fourth series, vol. xxxvi., p. 348.

mine the independent integrals X_1 and X_2; the integral of the equation in the original form being then given by equation (5) of the preceding article.

Substituting in equation (1)

$$u = \Sigma_0^\infty A_r x^{m+2qr},$$

we have

$$\Sigma_0^\infty \left[(m+2qr)(m+2qr-1)A_r x^{m+2qr-2} - a^2 A_r x^{m+2qr+2q-2} \right] = 0.$$

Equating to zero the coefficient of $x^{m+2qr-2}$, we have

$$(m+2qr)(m+2qr-1)A_r = a^2 A_{r-1},$$

and, when $r=0$,

$$m(m-1)A_0 = 0,$$

whence $m=0$ or $m=1$. Taking $m=0$, we obtain the integral

$$u_1 = 1 + \frac{a^2}{2q(2q-1)} x^{2q} + \frac{a^4}{2q \cdot 4q(2q-1)(4q-1)} x^{4q} + \ldots,$$

and, taking $m=1$,

$$u_2 = x \left(1 + \frac{a^2}{2q(2q+1)} x^{2q} + \frac{a^4}{2q \cdot 4q(2q+1)(4q+1)} x^{4q} + \ldots \right).$$

207. The integrals u_1 and u_2 are in no case finite series, nor do they fulfil the condition given in Art. 202 for expression in finite form, since in the notation there employed α and β are infinite. Let us, however, apply the transformation,

$$u = e^{\alpha x^m} v,$$

considered in Art. 154, and, if possible, determine α and m in such a manner that the transformed equation shall still be binomial. The equation for v is

§ XVII.] *INTEGRALS IN SERIES.* 227

$$\frac{d^2v}{dx^2} + 2max^{m-1}\frac{dv}{dx} + [m^2a^2x^{2m-2} + m(m-1)ax^{m-2} - a^2x^{2q-2}]v = 0,$$

which, it will be noticed, becomes a binomial equation if we put $m = q$ and $m^2a^2 = a^2$, whence $a = \pm\frac{a}{q}$. Hence we may put

$$u = e^{\frac{a}{q}x^q}v,$$

the transformed equation then being

$$\frac{d^2v}{dx^2} + 2ax^{q-1}\frac{dv}{dx} + a(q-1)x^{q-2}v = 0; \quad \ldots \quad (2)$$

and in the results we may change the sign of a, as is indeed evident from the form of equation (1).

208. Putting in equation (2) $v = \Sigma_0^\infty A_r x^{m+rq}$, we have

$$\Sigma_0^\infty [(m+rq)(m+rq-1)A_r x^{m+rq-2}$$
$$+ 2a(m+rq)A_r x^{m+rq+q-2} + a(q-1)A_r x^{m+rq+q-2}] = 0;$$

whence, equating to zero the coefficient of x^{m+rq-2}, we derive

$$(m+rq)(m+rq-1)A_r + a(2m + 2rq - q - 1)A_{r-1} = 0,$$

and, when $r = 0$,

$$m(m-1)A_0 = 0,$$

whence $m = 0$ or $m = 1$. Taking $m = 0$, we have

$$A_r = -\frac{(2r-1)q-1}{rq(rq-1)}aA_{r-1},$$

which gives, for the first integral of equation (2),

$$v_1 = 1 - \frac{q-1}{q(q-1)}ax^q + \frac{(q-1)(3q-1)}{q\cdot 2q(q-1)(2q-1)}a^2x^{2q} - \ldots;$$

and, taking $m = 1$,

$$B_r = -\frac{(2r-1)q+1}{rq(rq+1)}aB_{r-1},$$

which gives, for the second integral,

$$v_2 = x\left(1 - \frac{q+1}{q(q+1)}ax^q + \frac{(q+1)(3q+1)}{q\cdot 2q(q+1)(2q+1)}a^2x^{2q} - \ldots\right).$$

We have thus two new integrals of the equation

$$\frac{d^2u}{dx^2} - a^2x^{2q-2}u = 0,$$

namely,

$$u_3 = e^{\frac{a}{q}x^q}\left(1 - \frac{q-1}{q(q-1)}ax^q + \frac{(q-1)(3q-1)}{q\cdot 2q(q-1)(2q-1)}a^2x^{2q} - \ldots\right),$$

and

$$u_4 = xe^{\frac{a}{q}x^q}\left(1 - \frac{q+1}{q(q+1)}ax^q + \frac{(q+1)(3q+1)}{q\cdot 2q(q+1)(2q+1)}a^2x^{2q} - \ldots\right).$$

Again, changing the sign of a, we have two other integrals, namely,

$$u_5 = e^{-\frac{a}{q}x^q}\left(1 + \frac{q-1}{q(q-1)}ax^q + \frac{(q-1)(3q-1)}{q\cdot 2q(q-1)(2q-1)}a^2x^{2q} + \ldots\right),$$

and

$$u_6 = xe^{-\frac{a}{q}x^q}\left(1 + \frac{q+1}{q(q+1)}ax^q + \frac{(q+1)(3q+1)}{q\cdot 2q(q+1)(2q+1)}a^2x^{2q} + \ldots\right).$$

Finite Solutions.

209. When q is the reciprocal of the positive odd integer $2k - 1$, a zero factor occurs in the numerator of the coefficient of x^{kq} and of every subsequent term in the series contained in u_3. Notwithstanding the fact that the same zero factor occurs in the denominator of the same or a subsequent term, we have then a finite solution, as explained in the foot-note on page 175. At the same time, u_5 gives a finite integral, which is simply the result of changing the sign of a in that derived from u_3. For example, if $q = 1$, u_3 gives the integral e^{ax}, and u_5 gives e^{-ax}, so that we may write the complete integral $u = Ae^{ax} + Be^{-ax}$. Again, when $q = \tfrac{1}{3}$, the equation in the linear form being

$$\frac{d^2u}{dx^2} - a^2x^{-\tfrac{4}{3}}u = 0,$$

the two finite integrals are

$$e^{3ax^{\tfrac{1}{3}}}(1 - .3ax^{\tfrac{1}{3}}), \quad \text{and} \quad e^{-3ax^{\tfrac{1}{3}}}(1 + 3ax^{\tfrac{1}{3}}).$$

Since $m = 2q - 2$, the equation of Riccati in its non-linear form (2), Art. 204, is in this case

$$\frac{dy}{dx} + y^2 = a^2x^{-\tfrac{4}{3}},$$

and, substituting in equation (5), Art. 205, the results just found, the complete integral of this equation is found to be

$$y = \frac{-3a^2x^{-\tfrac{1}{3}}(e^{3ax^{\tfrac{1}{3}}} + ce^{-3ax^{\tfrac{1}{3}}})}{(1 - 3ax^{\tfrac{1}{3}})e^{3ax^{\tfrac{1}{3}}} + c(1 + 3ax^{\tfrac{1}{3}})e^{-3ax^{\tfrac{1}{3}}}}.$$

We may, if we please, express this solution in a logarithmic form; for, solving for c, we have

$$e^{6ax^{\tfrac{1}{3}}}\frac{3ax^{\tfrac{1}{3}}y - y - 3a^2x^{-\tfrac{1}{3}}}{3ax^{\tfrac{1}{3}}y + y + 3a^2x^{-\tfrac{1}{3}}} = c,$$

whence

$$\log \frac{3ax^{\frac{2}{3}}y - x^{\frac{1}{3}}y - 3a^2}{3ax^{\frac{2}{3}}y + x^{\frac{1}{3}}y + 3a^2} + 6ax^{\frac{1}{3}} = C.$$

210. In like manner, if q is the reciprocal of a negative odd integer, u_4 and u_6 give finite independent integrals. Hence we have a complete solution in finite terms, when q is of the form $\frac{1}{2k+1}$, where k is any integer. Substituting this expression in $m = 2q - 2$, we have

$$m = \frac{2}{2k+1} - 2 = \frac{-4k}{2k+1},$$

where k is any integer. Changing the sign of k, this expression becomes $\frac{-4k}{2k-1}$; so that the condition of integrability in finite form for equation (2), Art. 204, is that m should be of the form

$$\frac{-4k}{2k \pm 1},$$

where k is zero or a positive integer.

Relations between the Six Integrals.

211. Since u_3, Art. 208, is an integral of equation (1), it must be of the form $Au_1 + Bu_2$, where u_1 and u_2 are the integrals given in Art. 206. Now

$$e^{\frac{a}{q}x^q} = 1 + \frac{a}{q}x^q + \frac{a^2}{q^2}\frac{x^{2q}}{2!} + \frac{a^3}{q^3}\frac{x^{3q}}{3!} + \cdots,$$

and u_3 is the product of this series and the series

$$v_1 = 1 - \frac{a}{q}x^q + \frac{a^2}{q^2}\frac{3q-1}{2q-1}\frac{x^{2q}}{2!} - \cdots.$$

This product is a series having for its first term unity, and proceeding by integral powers of x^q. But u_1 is a series involving these powers, while in general u_2 contains none of these powers. It follows that, putting $u_3 = Au_1 + Bu_2$, we must have $B = 0$ and $A = 1$; that is, $u_3 = u_1$, the odd powers of x^q vanishing in the product. In like manner we can show that $u_5 = u_1$, and that $u_6 = u_4 = u_2$.

212. It thus appears that u_3 and u_5 are not independent integrals, but merely different expressions for the same function; nevertheless we have, in Art. 209, derived from them independent integrals in the case where they furnish finite expressions, namely, when q is the reciprocal of a positive odd integer. The explanation is that the finite expressions are *not* the actual values of u_3 and u_5, in these cases, but the results of rejecting from the series involved the infinite series of terms in which the vanishing factor occurs in the denominator as well as in the numerator of each coefficient. The rejected part of the series v_1, Art. 208, is a multiple of the series v_2; so that the finite expression, which we may denote by $[u_3]$, differs from u_3 by a multiple of u_4.

The expansion of the complete product u_3 is still the series u_1, consisting of even powers only of x^q; but that of the product $[u_3]$ contains also odd powers of x^q. These odd powers are accompanied by odd powers of a; hence, since $[u_5]$ is the result of changing the sign of a in $[u_3]$, it is evident that we shall have

$$u_1 = \tfrac{1}{2}[u_3] + \tfrac{1}{2}[u_5].$$

For example, when $q = 1$, $[u_3] = e^{ax}$, and $[u_5] = e^{-ax}$, of which the expansions contain both even and odd powers of x, but u_1 is the even function $\cosh x = \tfrac{1}{2}(e^{ax} + e^{-ax})$.

In like manner, when q is the reciprocal of a negative odd integer, we have the independent integrals $[u_4]$ and $[u_6]$, and

$$u_2 = \tfrac{1}{2}[u_4] + \tfrac{1}{2}[u_6].$$

Transformations of Riccati's Equation.

213. Certain important differential equations may be derived by transforming Riccati's equation,

$$\frac{d^2u}{dx^2} - a^2x^{2q-2}u = 0. \quad \ldots \quad (1)$$

For this purpose it is convenient to use the ϑ-form of the equation, namely,

$$\vartheta(\vartheta - 1)u - a^2x^{2q}u = 0. \quad \ldots \quad (2)$$

Let us first change the independent variable from x to z, where

$$z = mx^q, \quad \text{whence} \quad \vartheta' = z\frac{d}{dz} = \frac{1}{q}\vartheta.$$

The result is

$$q\vartheta'(q\vartheta' - 1)u - a^2\frac{z^2}{m^2}u = 0;$$

putting $m = \frac{1}{q}$, and writing ϑ and x in place of ϑ' and z, this becomes

$$\vartheta(\vartheta - m)u - a^2x^2u = 0, \quad \ldots \quad (3)$$

which in the ordinary notation is

$$\frac{d^2u}{dx^2} - \frac{m-1}{x}\frac{du}{dx} - a^2u = 0. \quad \ldots \quad (4)$$

Hence, putting $q = \frac{1}{m}$, and writing $\frac{x}{m}$ in place of x^q in the six values of u given in Arts. 206 and 208, we have the following six integrals of equation (4),

$$u_1 = 1 - \frac{a^2}{m-2}\frac{x^2}{2} + \frac{a^4}{(m-2)(m-4)}\frac{x^4}{2^2 2!} - \cdots,$$

$$u_2 = x^m\left(1 + \frac{a^2}{m+2}\frac{x^2}{2} + \frac{a^4}{(m+2)(m+4)}\frac{x^4}{2^2 2!} + \cdots\right),$$

$$u_3 = e^{ax}\left(1 - \frac{m-1}{m-1}ax + \frac{(m-1)(m-3)}{(m-1)(m-2)}\frac{a^2x^2}{2!} - \cdots\right),$$

$$u_4 = e^{ax}x^m\left(1 - \frac{m+1}{m+1}ax + \frac{(m+1)(m+3)}{(m+1)(m+2)}\frac{a^2x^2}{2!} - \cdots\right),$$

$$u_5 = e^{-ax}\left(1 + \frac{m-1}{m-1}ax + \frac{(m-1)(m-3)}{(m-1)(m-2)}\frac{a^2x^2}{2!} + \cdots\right),$$

$$u_6 = e^{-ax}x^m\left(1 + \frac{m+1}{m+1}ax + \frac{(m+1)(m+3)}{(m+1)(m+2)}\frac{a^2x^2}{2!} + \cdots\right).$$

The factor m^m has been omitted in writing u_2, u_4, and u_6, but we still have $u_1 = u_3 = u_5$, and $u_2 = u_4 = u_6$.

Equation (4) is integrable in finite terms when m is an odd integer, the complete integral being $A[u_3] + B[u_5]$ when m is positive, and $A[u_4] + B[u_6]$ when m is negative.

214. If in equation (3) we put $m = 2p + 1$, and make the transformation

$$u = x^p v,$$

we have, since $\vartheta x^p V = p x^p V + x^p \vartheta V = x^p(\vartheta + p)V$,

$$(\vartheta + p)(\vartheta - p - 1)v - a^2x^2v = 0,$$

which in the ordinary notation is

$$\frac{d^2v}{dx^2} - a^2v = \frac{p(p+1)}{x^2}v.$$

This equation is integrable in finite terms when p is an integer.* The case in which $p = 2$ occurs in investigations concerning the figure of the earth.

---

* See the memoir "On RICCATI's Equation and its Transformations, and on some Definite Integrals which satisfy them," by J. W. L. Glaisher, *Philosophical Transactions* for 1881, in which the six integrals of this equation are deduced directly, and those of the equations treated in the preceding articles are derived from them.

Bessel's Equation.

215. If, in equation (3), Art. 213, we put $m = 2n$ and $a^2 = -1$, and make the transformation $u = x^n y$, the result is

$$(\vartheta^2 - n^2)y + x^2 y = 0, \quad \ldots \ldots \ldots (1)$$

or, in the ordinary notation,

$$x^2 \frac{d^2y}{dx^2} + x \frac{dy}{dx} + (x^2 - n^2)y = 0, \quad \ldots \ldots (2)$$

which is known as *Bessel's Equation*. Making the substitutions in the values of u_2 and u_1, Art. 213, and denoting the corresponding integrals of Bessel's equation by y_n and y_{-n}, we have

$$y_n = x^n \left(1 - \frac{1}{n+1} \frac{x^2}{2^2} + \frac{1}{(n+1)(n+2)} \frac{x^4}{2^4 \cdot 2!} - \cdots\right),$$

$$y_{-n} = x^{-n} \left(1 + \frac{1}{n-1} \frac{x^2}{2^2} + \frac{1}{(n-1)(n-2)} \frac{x^4}{2^4 \cdot 2!} + \cdots\right).$$

It will be noticed that either of these integrals may be obtained from the other by changing the sign of n, which we are at liberty to do by virtue of the form of the differential equation.

216. The integrals corresponding to the other four values of u in Art. 213 are imaginary in form. Making the substitutions in the value of u_4, we may write, since $u_4 = u_2 = x^n y_n$,

$$y_n = x^n (\cos x + i \sin x)(P_n - i Q_n),$$

in which

$$P_n = 1 - \frac{(2n+1)(2n+3)}{(2n+1)(2n+2)} \frac{x^2}{2!} + \cdots,$$

$$Q_n = \frac{2n+1}{2n+1} x - \frac{(2n+1)(2n+3)(2n+5)}{(2n+1)(2n+2)(2n+3)} \frac{x^3}{3!} + \cdots.$$

The value of y_n derived from u_6 is the same thing with the sign of i changed; hence we infer that

$$y_n = x^n(P_n \cos x + Q_n \sin x),$$

and also that

$$P_n \sin x - Q_n \cos x = 0.^*$$

Changing the sign of n, the other integral of Bessel's equation may, in like manner, be written in the form

$$y_{-n} = x^{-n}(P_{-n} \cos x + Q_{-n} \sin x),$$

where

$$P_{-n} = 1 - \frac{(2n-1)(2n-3)}{(2n-1)(2n-2)}\frac{x^2}{2!} + \cdots,$$

$$Q_{-n} = \frac{2n-1}{2n-1}x - \frac{(2n-1)(2n-3)(2n-5)}{(2n-1)(2n-2)(2n-3)}\frac{x^3}{3!} + \cdots.$$

Finite Solutions.

217. The case in which Bessel's equation admits of finite solution is that in which n is one-half of an odd integer. Taking n to be positive, the series P_{-n} and Q_{-n} contain, in this case, terms whose coefficients have zero factors in the numerators. Denoting by $[P_{-n}]$ and $[Q_{-n}]$ the finite series preceding these terms, we have, as explained in Art. 212, an integral $[y_{-n}]$ in finite form, but differing in value from y_{-n}. Thus

---

* The resulting value of $\tan x$ may be written

$$\tan x = \frac{x - \frac{m+5}{m+2}\frac{x^3}{3!} + \frac{(m+7)(m+9)}{(m+2)(m+4)}\frac{x^5}{5!} - \cdots}{1 - \frac{m+3}{m+2}\frac{x^2}{2!} + \frac{(m+5)(m+7)}{(m+2)(m+4)}\frac{x^4}{4!} - \cdots},$$

in which m may have any value.

$$[y_{-n}] = x^{-n}(\cos x + i \sin x)\{[P_{-n}] - i[Q_{-n}]\}$$
$$= x^{-n}\{\cos x\, [P_{-n}] + \sin x[Q_{-n}]\}$$
$$+ ix^{-n}\{\sin x[P_{-n}] - \cos x[Q_{-n}]\},$$

in which the coefficient of i does *not* vanish, as it does in Art. 216. If we substitute this expression in the differential equation, it is evident that the real and imaginary parts of the result must separately vanish, so that we have the two real integrals

$$\eta_1 = x^{-n}\{\cos x\,[P_{-n}] + \sin x[Q_{-n}]\},$$

and

$$\eta_2 = x^{-n}\{\sin x\,[P_{-n}] - \cos x[Q_{-n}]\}.$$

The complete integral may therefore, in this case, be written

$$y = Cx^{-n}\{[P_{-n}]\cos(x+a) + [Q_{-n}]\sin(x+a)\},$$

where C and a are the constants of integration.

218. Comparing the integrals η_1 and η_2 with y_{-n} and y_n, Art. 215, it is evident that, since $\cos x[P_{-n}] + \sin x[Q_{-n}]$ is an even function, and $\sin x[P_{-n}] - \cos x[Q_{-n}]$ is an odd function, the development of η_1 contains only the powers of x which occur in y_{-n}, and η_2 only those which occur in y_n. Moreover, the first coefficient in η_1 is unity. It follows that $\eta_1 = y_{-n}$, and that η_2 is the product of y_n by a constant.*

---

* To find this constant, we notice that the part of the series $P_{-n} - iQ_{-n}$, which is rejected from the value of y_{-n}, when we use the finite expressions, as in Art. 217, commences with the term containing x^{2n}. Denoting the coefficient of this term by A, the rejected part of y_{-n} is Ay_n. Thus

$$y_{-n} = [y_{-n}] + Ay_n = \eta_1 + i\eta_2 + Ay_n.$$

But we have shown that $\eta_1 = y_{-n}$; hence $\eta_2 = -\dfrac{A}{i}y_n$, where A is the coefficient of x^{2n} in $P_{-n} - iQ_{-n}$, that is, in $-iQ_{-n}$, Art. 216, since $2n$ is an odd integer. Thus

$$\eta_2 = \frac{2^{2n-1}[(n-\tfrac{1}{2})!]^2}{(2n-1)!(2n)!}y_n.$$

The Besselian Function.

219. If, when n is a positive integer, we multiply y_n, Art. 215, by the constant $\dfrac{1}{2^n n!}$, the resulting integral of Bessel's equation is known as the *Besselian function* of the nth order, and is denoted by J_n. Thus

$$J_n = \frac{x^n}{2^n n!}\left(1 - \frac{1}{n+1}\frac{x^2}{2^2} + \frac{1}{(n+1)(n+2)}\frac{x^4}{2^4 2!} - \cdots\right)$$

$$= \sum_0^\infty \frac{(-1)^r}{(n+r)!\, r!}\left(\frac{x}{2}\right)^{n+2r}.$$

More generally, for all values of n we may write

$$J_n = \frac{x^n}{2^n \Gamma(n+1)}\left(1 - \frac{1}{n+1}\frac{x^2}{2^2} + \frac{1}{(n+1)(n+2)}\frac{x^4}{2^4 2!} - \cdots\right)$$

$$= \sum_0^\infty \frac{(-1)^r}{\Gamma(n+1+r)\, r!}\left(\frac{x}{2}\right)^{n+2r},$$

and then, in general, the complete integral of Bessel's equation is

$$y = AJ_n + BJ_{-n},$$

where J_{-n} is the same function of $-n$ that J_n is of n. It is to be noticed, however, that the factor which converts the series y_{-n} to J_{-n} is zero in value when n is a positive integer.

Substituting the values of η_2 and y_n, Arts. 217 and 215, we have, for the development of the odd function $\sin x[P_{-n}] - \cos x[Q_{-n}]$,

$$\frac{2^{2n-1}[(n-\frac{1}{2})!]^2}{(2n-1)!\,(2n)!} x^{2n}\left(1 - \frac{x^2}{2(2n+2)} + \frac{x^4}{2 \cdot 4(2n+2)(2n+4)} - \cdots\right).$$

The series in this case contains infinite terms which are thus rendered finite, while the finite terms preceding that which contains x^{2n} are made to vanish. The result is that, when n is an integer,

$$J_{-n} = (-1)^n J_n,$$

and the expression $AJ_n + BJ_{-n}$ fails to represent the complete integral. The second integral in this case takes the logarithmic form, and is found in Art. 221.

220. The expression for y_n, given in Art. 216, shows that

$$J_n = \frac{x^n}{2^n \Gamma(n+1)} (P_n \cos x + Q_n \sin x),^*$$

where

$$P_n = 1 - \frac{2n+3}{2n+2} \frac{x^2}{2!} + \frac{(2n+5)(2n+7)}{(2n+2)(2n+4)} \frac{x^4}{4!} - \cdots,$$

$$Q_n = x - \frac{2n+5}{2n+2} \frac{x^3}{3!} + \frac{(2n+7)(2n+9)}{(2n+2)(2n+4)} \frac{x^5}{5!} - \cdots.$$

---

* Finite expressions for J_n and J_{-n} exist when n is of the form $p + \frac{1}{2}$, p being an integer. These are multiples of η_2 and η_1, Art. 217, respectively. Substituting in the numerical factors the values of the corresponding Gamma functions, which are

$$\Gamma(p + \tfrac{3}{2}) = (p + \tfrac{1}{2}) \Gamma(p + \tfrac{1}{2}) = \frac{(2p+1)!}{2^{2p+1} p!} \sqrt{\pi},$$

and

$$\Gamma(\tfrac{1}{2} - p) = (-1)^p \frac{2^{2p} p!}{(2p)!} \sqrt{\pi};$$

and, taking account also, in the case of $J_{p+\frac{1}{2}}$, of the factor found in the preceding foot-note, we find

$$J_{p+\frac{1}{2}} = \frac{(2p)!}{2^{p-\frac{1}{2}} p! \sqrt{\pi}} \frac{\sin x [P_{-(p+\frac{1}{2})}] - \cos x [Q_{-(p+\frac{1}{2})}]}{x^{p+\frac{1}{2}}},$$

and

$$J_{-(p+\frac{1}{2})} = (-1)^p \frac{(2p)!}{2^{p-\frac{1}{2}} p! \sqrt{\pi}} \frac{\cos x [P_{-(p+\frac{1}{2})}] + \sin x [Q_{-(p+\frac{1}{2})}]}{x^{p+\frac{1}{2}}}.$$

The Besselian Function of the Second Kind.

221. The second integral, when n is an integer, may be found by the process employed in Arts. 175 and 176, and in similar cases. Thus, changing equation (1), Art. 215, to

$$(\vartheta - n)(\vartheta + n - h)y + x^2 y = 0,$$

and putting $y = \Sigma_{-\infty}^{0} A_r x^{m+2r}$, the relation between consecutive coefficients is

$$A_r = -\frac{A_{r-1}}{(m+2r-n)(m+2r+n-h')},$$

where h' is put for h. Making $m = n$ and $m = -n + h$ successively, we have the integrals

$$y_n = x^n \left(1 - \frac{x^2}{2(2n+2-h')} + \frac{x^4}{2.4(2n+2-h')(2n+4-h')} - \ldots \right),$$

and

$$y_{-n} = x^{-n+h}\left[1 + \frac{x^2}{(2+h-h')(2n-2-h)} + \ldots \right.$$

$$+ \frac{x^{2n}}{(2+h-h')\ldots(2n+h-h')(2n-2-h)\ldots(-h)}$$

$$\left. \times \left(1 - \frac{x^2}{(2n+2+h-h')(2+h)} + \ldots \right) \right.$$

Denoting the product of x^n and the series last written by $\psi(h)$, we have $\psi(0) = y_n$, and the complete integral $y = A_0 y_n + B_0 y_{-n}$ may be written

$$y = A_0 y_n + B_0 T + \frac{B}{h}(1 + h \log x + \ldots)[y_n + h \psi'(0) + \ldots],$$

where, when $h = 0$,

$$B = -\frac{B_0}{2.4 \ldots 2n(2n-2)(2n-4) \ldots 2} = -\frac{B_0}{2^{2n-1} n! (n-1)!},$$

and T denotes the aggregate of terms in y_{-n}, which remain finite when $h = 0$. We have therefore

$$y = A y_n + B_0 T + B y_n \log x + B \psi'(0) + \ldots,$$

and may take as the second integral, when $h = 0$,

$$y_n \log x - 2^{2n-1} n! (n-1)! T + \psi'(0).$$

If this expression be divided by $2^n n!$, the first term becomes $J_n \log x$; denoting the quotient by Y_n, and developing $\psi'(0)$ as in Art. 173, we have

$$Y_n = J_n \log x - 2^{n-1}(n-1)! x^{-n} \left[1 + \frac{1}{n-1} \frac{x^2}{2^2} \right.$$

$$\left. + \frac{1}{(n-1)(n-2)} \frac{x^4}{2^4 \cdot 2!} + \ldots + \frac{1}{(n-1)!} \frac{x^{2n-2}}{2^{2n-2}(n-1)!} \right]$$

$$+ \frac{x^n}{2^{n+1} n!} \left[\frac{1}{1 \cdot (n+1)} \left(1 + \frac{1}{n+1} \right) \frac{x^2}{2^2} \right.$$

$$\left. - \frac{1}{1 \cdot 2 \cdot (n+1)(n+2)} \left(1 + \frac{1}{2} + \frac{1}{n+1} + \frac{1}{n+2} \right) \frac{x^4}{2^4} + \ldots \right],$$

and the complete integral of Bessel's equation, when n is an integer may be written

$$y = A J_n + B Y_n.$$

The integral Y_n is called the Besselian function of the second kind.*

Legendre's Equation.

222. The equation

$$(1 - x^2)\frac{d^2y}{dx^2} - 2x\frac{dy}{dx} + n(n+1)y = 0, \quad \ldots \quad (1)$$

or, as it may be written,

$$\frac{d}{dx}\left\{(1-x^2)\frac{dy}{dx}\right\} + n(n+1)y = 0,$$

is known as *Legendre's Equation*, because, when n is an integer, it is the differential equation satisfied by the nth member of a set of rational integral functions of x known as the Legendrean Coefficients.† Particular interest, therefore, attaches to the case in which n is a positive integer; and it is to be noticed

---

* The properties of the Besselian functions are discussed in Lommel's "Studien über die Bessel'schen Functionen," Leipzig, 1868; Todhunter's "Treatise on Laplace's Functions, Lamé's Functions, and Bessel's Functions," London, 1875, etc.

† The Legendrean Coefficient of the nth order is the coefficient of a^n in the expansion in ascending powers of a of the expression

$$V = \frac{1}{\sqrt{(1 - 2ax + a^2)}},$$

and is denoted by $P_n(x)$, or simply by P_n. It is readily shown that

$$\frac{d}{dx}\left\{(1-x^2)\frac{dV}{dx}\right\} + \frac{d}{da}\left\{a^2\frac{dV}{da}\right\} = 0;$$

whence, substituting $V = \Sigma_0^\infty a^n P_n$ and equating to zero the coefficient of a^n, we find

$$\frac{d}{dx}\left\{(1-x^2)\frac{dP_n}{dx}\right\} + n(n+1)P_n = 0.$$

When $x = 1$, $V = \dfrac{1}{1-a} = 1 + a + a^2 + \ldots$; hence $P_n(1) = 1$ for all values of n.

that this includes the case in which n is a negative integer; for, if in that case we put $-n = n' + 1$, whence $-(n+1) = n'$, we shall have an equation of the same form in which n' is zero or a positive integer.

223. When written in the ϑ-form, Legendre's equation is

$$\vartheta(\vartheta - 1)y - x^2(\vartheta - n)(\vartheta + n + 1)y = 0,$$

a binomial equation in which both terms are of the second degree in ϑ. Hence the equation may be solved in series proceeding either by ascending or descending powers of x. Putting $y = \Sigma_0^\infty A_r x^{m+2r}$, we have, for the integrals in ascending series,

$$y_1 = 1 - n(n+1)\frac{x^2}{2!} + n(n-2)(n+1)(n+3)\frac{x^4}{4!} - \ldots,$$

and

$$y_2 = x - (n-1)(n+2)\frac{x^3}{3!} + (n-1)(n-3)(n+2)(n+4)\frac{x^5}{5!} - \ldots.$$

Again, writing the equation in the form

$$(\vartheta - n)(\vartheta + n + 1)y - x^{-2}\vartheta(\vartheta - 1)y = 0,$$

and putting $y = \Sigma_0^\infty A_r x^{m-2r}$, we have the integrals in descending series

$$y_3 = x^n\left(1 - \frac{n(n-1)}{2(2n-1)}x^{-2} + \frac{n(n-1)(n-2)(n-3)}{2 \cdot 4(2n-1)(2n-3)}x^{-4} - \ldots\right),$$

and

$$y_4 = x^{-n-1}\left(1 + \frac{(n+1)(n+2)}{2(2n+3)}x^{-2}\right.$$
$$\left. + \frac{(n+1)(n+2)(n+3)(n+4)}{2 \cdot 4(2n+3)(2n+5)}x^{-4} + \ldots\right).$$

The Legendrean Coefficients.

224. When n is a positive integer, y_1 or y_2 is a finite series according as n is even or odd; and in either case y_3 is a finite expression, differing from y_1 or y_2 only by a constant factor. If y_3 be multiplied by the constant

$$\frac{(2n-1)(2n-3)\ldots 1}{n!} \quad \text{or} \quad \frac{(2n)!}{2^n(n!)^2},$$

the resulting integral is the Legendrean coefficient of the nth order, which is denoted by P_n. By the cancellation of common factors in the numerators and denominators of the coefficients, the successive values of P_n may be written as follows:—

$$P_0 = 1, \qquad P_2 = \frac{3}{2}x^2 - \frac{1}{2},$$

$$P_1 = x, \qquad P_3 = \frac{5}{2}x^3 - \frac{3}{2}x,$$

$$P_4 = \frac{7 \cdot 5}{4 \cdot 2}x^4 - 2\frac{5 \cdot 3}{4 \cdot 2}x^2 + \frac{3 \cdot 1}{4 \cdot 2},$$

$$P_5 = \frac{9 \cdot 7}{4 \cdot 2}x^5 - 2\frac{7 \cdot 5}{4 \cdot 2}x^3 + \frac{5 \cdot 3}{4 \cdot 2}x,$$

$$P_6 = \frac{11 \cdot 9 \cdot 7}{6 \cdot 4 \cdot 2}x^6 - 3\frac{9 \cdot 7 \cdot 5}{6 \cdot 4 \cdot 2}x^4 + 3\frac{7 \cdot 5 \cdot 3}{6 \cdot 4 \cdot 2}x^2 - \frac{5 \cdot 3 \cdot 1}{6 \cdot 4 \cdot 2},$$

$$\ldots \quad \ldots \quad \ldots \quad \ldots \quad \ldots \quad \ldots,$$

in which the law of formation of the coefficients is obvious.*

---

* The constant is so taken that the definition of P_n given above agrees with that given in the preceding foot-note. For, putting $x = 1$, and forming the differences of the successive fractions which in the expressions last written are multiplied by the binomial coefficients, it is readily shown that $P_n(1) = 1$, for all values of n.

The Second Integral when n is an Integer.

225. When n is an integer, the second integral of Legendre's equation admits of expression in a finite form.

Assume
$$y = uP_n - v, \qquad \ldots \ldots (1)$$

where u and v are functions of x. By substitution in equation (1), Art. 222, we have

$$u\left\{(1-x^2)\frac{d^2P_n}{dx} - 2x\frac{dP_n}{dx} + n(n+1)P_n\right\} + 2(1-x^2)\frac{du}{dx}\frac{dP_n}{dx}$$

$$+ P_n\left\{(1-x^2)\frac{d^2u}{dx^2} - 2x\frac{du}{dx}\right\} - (1-x^2)\frac{d^2v}{dx^2} + 2x\frac{dv}{dx} - n(n+1)v = 0,$$

in which the coefficient of u vanishes, because P_n is an integral, and that of P_n will vanish if u be so taken that

$$(1-x^2)\frac{d^2u}{dx^2} - 2x\frac{du}{dx} = 0.$$

This condition is satisfied if we take $(1-x^2)\dfrac{du}{dx} = 1$, whence

$$u = \tfrac{1}{2}\log\frac{x+1}{x-1}; \qquad \ldots \ldots (2)$$

the equation then becomes

$$(1-x^2)\frac{d^2v}{dx^2} - 2x\frac{dv}{dx} + n(n+1)v = 2\frac{dP_n}{dx}, \quad \ldots (3)$$

and we shall have a solution of Legendre's equation in the assumed form (1), if v is determined as a particular integral of this equation.

Now, since P_n is a rational and integral algebraic function of the nth degree, the second member of equation (3) is an algebraic function of the $(n-1)$th degree; hence the particular integral required is the sum of those of several equations of the form

$$(1-x^2)\frac{d^2y}{dx^2} - 2x\frac{dy}{dx} + n(n+1)y = ax^p, \quad \ldots \quad (4)$$

in which p is a positive integer less than n. Solving equation (4) in descending series, the particular integral is

$$Y = -\frac{ax^p}{(p-n)(p+n+1)}\left(1 + \frac{p(p-1)}{(p+n-1)(p-n-2)}x^{-2}\right.$$
$$\left. + \frac{p(p-1)(p-2)(p-3)}{(p+n-1)(p+n-3)(p-n-2)(p-n-4)}x^{-4} + \ldots\right),$$

which, when p is an integer, is a finite series containing no negative powers of x. Thus the particular integral of equation (4) is an algebraic function of x of the pth degree, and that of equation (3) is an algebraic function of the $(n-1)$th degree. Denoting this function by R_n, we have therefore an integral of Legendre's equation of the form

$$Q_n = \tfrac{1}{2}P_n \log\frac{x+1}{x-1} - R_n. \quad \ldots \ldots \quad (5)$$

226. Since

$$\tfrac{1}{2}\log\frac{x+1}{x-1} = \frac{1}{x} + \frac{1}{3x^3} + \frac{1}{5x^5} + \ldots,$$

the product $\tfrac{1}{2}P_n \log\dfrac{x+1}{x-1}$, when developed in descending series, commences with the term containing x^{n-1}; and as R_n contains no terms of higher degree, the development of Q_n cannot contain x^n. It follows that, putting $Q_n = Ay_3 + By_4$ where y_3 and y_4 are the integrals in descending series, Art. 223, we must have $Q_n = By_4$.* But y_4 commences with the term x^{-n-1}; we therefore infer that in the product above mentioned the terms with positive exponents are the same as those of R_n, and are cancelled thereby in the development of Q_n, while the terms with negative exponents vanish until we reach the term Bx^{-n-1}. The formation of the required terms of this product affords a ready method of calculating R_n.†

---

* To determine the value of B, we notice that equation (3), Art. 147, gives, for the relation between the integrals P_n and Q_n of Legendre's equation,

$$P_n \frac{dQ_n}{dx} - Q_n \frac{dP_n}{dx} = \frac{A}{1-x^2}, \quad \cdots \cdots (1)$$

where A is a definite constant. Substituting from equation (5), this gives

$$P_n^2 + (x^2 - 1)\left[P_n \frac{dR_n}{dx} - R_n \frac{dP_n}{dx}\right] = A.$$

Putting $x = 1$, we have $A = 1$, because $P_n(1) = 1$, and P_n and R_n being rational integral functions, the quantity in brackets does not become infinite. Now, from Art. 224, $P_n = \frac{(2n)!}{2^n(n!)^2} y_3$; substituting this value, and putting $A = 1$, $Q_n = By_4$, equation (1) becomes

$$\frac{B(2n)!}{2^n(n!)^2}\left(y_3 \frac{dy_4}{dx} - y_4 \frac{dy_3}{dx}\right) = \frac{1}{1-x^2}.$$

Developing both members in descending powers, and comparing the first terms, we have

$$\frac{B(2n)!}{2^n(n!)^2} x^{-2}(-n-1-n) = -x^{-2},$$

whence

$$B = \frac{2^n(n!)^2}{(2n+1)!};$$

that is

$$Q_n = \frac{2^n(n!)^2}{(2n+1)!} y_4.$$

† The Legendrean coefficients are sometimes called zonal harmonics, the term spherical harmonics (in French and German treatises *fonctions sphériques* and *Kugelfunctionen*) being applied to a more general class of functions which include them. The function Q_n is the zonal harmonic of the second kind. Discussions of the properties of the functions P_n and Q_n will be found in Todhunter's Treatise "On Laplace's Functions, Lamé's Functions, and Bessel's Functions," London, 1875; Ferrers' "Spherical Harmonics," London, 1877; Heine's "Handbuch der Kugelfunctionen," Berlin, 1878; etc.

Examples XVII.

Solve the following differential equations :—

1. $\dfrac{dy}{dx} + y^2 = \dfrac{a^2}{x^4}$, $\quad y = \dfrac{(x-a)e^{ax^{-1}} + c(x+a)e^{-ax^{-1}}}{x^2(e^{ax^{-1}} + ce^{-ax^{-1}})}$.

2. $\dfrac{d^2u}{dx^2} - a^2 x^{-1} u = 0$,

    $u = Axe^{-3ax^{-\frac{1}{2}}}(1 + 3ax^{-\frac{1}{2}}) + Bxe^{3ax^{-\frac{1}{2}}}(1 - 3ax^{-\frac{1}{2}})$.

3. $\dfrac{d^2u}{dx^2} - \dfrac{2}{x}\dfrac{du}{dx} - a^2 u = 0$, $\quad u = Ae^{ax}(1-ax) + Be^{-ax}(1+ax)$.

4. $\dfrac{d^2u}{dx^2} + \dfrac{2}{x}\dfrac{du}{dx} - a^2 u = 0$, $\quad u = x^{-1}(Ae^{ax} + Be^{-ax})$.

5. $\dfrac{d^2u}{dx^2} + \dfrac{2}{x}\dfrac{du}{dx} + a^2 u = 0$, $\quad u = C\dfrac{\cos(x-a)}{x}$.

6. $\dfrac{d^2u}{dx^2} + \dfrac{4}{x}\dfrac{du}{dx} - a^2 u = 0$,

    $u = Ax^{-3}e^{ax}(1 - ax) + Bx^{-3}e^{-ax}(1 + ax)$.

7. $\dfrac{d^2u}{dx^2} + \dfrac{4}{x}\dfrac{du}{dx} + a^2 u = 0$,

    $u = Ax^{-3}(\cos ax + ax \sin ax) + Bx^{-3}(\sin ax - ax \cos ax)$.

8. $\dfrac{d^2y}{dx^2} - a^2 y = \dfrac{6y}{x^2}$,

    $y = Ax^{-2}e^{ax}(1 - ax + \tfrac{1}{3}a^2x^2) + Bx^{-2}e^{-ax}(1 + ax + \tfrac{1}{3}a^2x^2)$.

9. $\dfrac{d^2y}{dx^2} + n^2 y = \dfrac{6y}{x^2}$,

    $y = Cx^{-2}[(3 - n^2x^2)\cos(nx + a) + 3nx \sin(nx + a)]$.

10. $x^2\dfrac{d^2y}{dx^2} + x\dfrac{dy}{dx} - (x^2 + \tfrac{1}{4})y = 0,\qquad y = \dfrac{Ae^x + Be^{-x}}{\sqrt{x}}.$

11. $x^2\dfrac{d^2y}{dx^2} + x\dfrac{dy}{dx} + \dfrac{4x^2 - 9a^2}{4a^2}y = 0,$

$$y = C\left(x^{-\frac{1}{2}}\cos\dfrac{x+a}{a} + x^{-\frac{1}{2}}\sin\dfrac{x+a}{a}\right).$$

12. $x^2\dfrac{d^2y}{dx^2} + x\dfrac{dy}{dx} + (x^2 - \tfrac{25}{4})y = 0,$

$$y = Cx^{-\frac{1}{2}}[(1 - \tfrac{1}{3}x^2)\cos(x+a) + x\sin(x+a)].$$

13. Show that, when q is the reciprocal of an odd integer, the integral of Riccati's equation,
$$\dfrac{d^2u}{dx^2} - a^2x^{2q-2}u = 0,$$
may be written in the form

$$u = Ax^{\frac{1}{2}(1-q)}e^{\frac{a}{q}x^q}\left[1 + \dfrac{q^2 - 1}{q}\dfrac{1}{8ax^q} + \dfrac{(q^2 - 1)(3^2q^2 - 1)}{q\cdot 2q}\left(\dfrac{1}{8ax^q}\right)^2 + \cdots\right]$$

$$+ Bx^{\frac{1}{2}(1-q)}e^{-\frac{a}{q}x^q}\left[1 - \dfrac{q^2 - 1}{q}\dfrac{1}{8ax^q} + \dfrac{(q^2 - 1)(3^2q^2 - 1)}{q\cdot 2q}\left(\dfrac{1}{8ax^q}\right)^2 - \cdots\right]$$

14. Show that for all values of n

$$e^{2x} = \dfrac{1 + x + \dfrac{n+2}{n+1}\dfrac{x^2}{2!} + \dfrac{(n+2)(n+4)}{(n+1)(n+2)}\dfrac{x^3}{3!} + \cdots}{1 - x + \dfrac{n+2}{n+1}\dfrac{x^2}{2!} - \dfrac{(n+2)(n+4)}{(n+1)(n+2)}\dfrac{x^3}{3!} + \cdots}.$$

15. Show that the complete integral of the equation
$$\dfrac{d^2y}{dx^2} + q\dfrac{dy}{dx} = \dfrac{2y}{x^2}$$
may be written in the form
$$xy = A(2 - qx) + Be^{-qx}(2 + qx).$$

16. If in Riccati's equation $a^2 = -1$, show that the integral may be expressed in Besselian functions.

$$u = \sqrt{x}\left[AJ_{\frac{1}{2q}}\left(\frac{1}{q}x^q\right) + BJ_{-\frac{1}{2q}}\left(\frac{1}{q}x^q\right)\right].$$

17. Reduce to Bessel's form the equation

$$x^2\frac{d^2y}{dx^2} + nx\frac{dy}{dx} + (b + cx^{2m})y = 0,$$

and show that its integral in Besselian functions is

$$y = x^{-\frac{1}{2}(n-1)}\left[AJ_\mu\left(c^{\frac{1}{2}}\frac{x^m}{m}\right) + BJ_{-\mu}\left(c^{\frac{1}{2}}\frac{x^m}{m}\right)\right],$$

where $\mu = \dfrac{\sqrt{[(n-1)^2 - 4b]}}{2m}$.

18. $\dfrac{d^2y}{dx^2} + ye^{2x} = n^2y,$ $\qquad y = AJ_n(e^x) + BJ_{-n}(e^x).$

19. $\dfrac{d^2y}{dx^2} + \dfrac{y}{4x} = 0,$ $\qquad y = x^{\frac{1}{2}}[AJ_1(x^{\frac{1}{2}}) + BY_1(x^{\frac{1}{2}})].$

20. $x\dfrac{d^2y}{dx^2} + \dfrac{dy}{dx} + y = 0,$ $\qquad y = AJ_0(2x^{\frac{1}{2}}) + BY_0(2x^{\frac{1}{2}}).$

21. $x\dfrac{d^2y}{dx^2} + 3\dfrac{dy}{dx} + 4x^3y = 0,$ $\qquad y = A\dfrac{\cos x^2}{x^2} + B\dfrac{\sin x^2}{x^2}.$

22. Putting $u = e^{a\sqrt{(x^2 + hx)}} = \sum_0^\infty P_r h^r$, show that

$$\frac{d^2u}{dx^2} - a^2u = \frac{h^2}{x^2}\frac{d^2u}{dh^2},$$

and thence that P_{p+1} is an integral of

$$\frac{d^2v}{dx^2} - a^2v = \frac{p(p+1)}{x^2}v.$$

23. P_m and P_n being Legendrean coefficients, show that
$$n(n+1)\int_{-1}^{1} P_n P_m dx = \int_{-1}^{1} (1-x^2)\frac{dP_n}{dx}\frac{dP_m}{dx}dx,$$
and thence that
$$\int_{-1}^{1} P_n P_m dx = 0,$$
except when $m = n$. Also show that, when $m + n$ is an even number, $\int_{0}^{1} P_n P_m dx = 0$, unless $m = n$.

CHAPTER X.

EQUATIONS INVOLVING MORE THAN TWO VARIABLES.

XVIII.

Determinate Systems of the First Order.

227. A system of n simultaneous equations between $n + 1$ variables and their differentials is a *determinate system* of the first order, because it serves to determine the ratios of the $n + 1$ differentials; so that, one of the variables being taken as independent, the others vary in a determinate manner, and may therefore be regarded as functions of the single independent variable.

A determinate system involving the variables $x, y, z, \ldots$ may be written in the symmetrical form

$$\frac{dx}{X} = \frac{dy}{Y} = \frac{dz}{Z} = \cdots,$$

in which $X, Y, Z, \ldots$ may be any functions of the variables.

228. When the system is put in this form, we may consider the several equations each of which involves two of the differentials; if one of these contains only the corresponding variables, it is an ordinary differential equation between two variables, and its integration gives us a relation between these two variables. This integral may be used to eliminate one of these variables from one of the other equations, and may thus enable us to obtain another equation containing only two variables; and finally, in this manner, n integral equations between the $n + 1$ variables. Given, for example, the system

$$\frac{dx}{y} = \frac{dy}{x} = \frac{dz}{z}, \quad \ldots \ldots \ldots (1)$$

in which the equation involving dx and dy is independent of z; integrating it, we have

$$x^2 - y^2 = a. \quad \ldots \ldots \ldots (2)$$

Employing this to eliminate x, the equation involving dy and dz becomes

$$\frac{dy}{\sqrt{(y^2 + a)}} = \frac{dz}{z},$$

and the integral of this is

$$y + \sqrt{(y^2 + a)} = bz. \quad \ldots \ldots \ldots (3)$$

The integral equations (2) and (3) containing two constants of integration constitute the complete solution of the given system.

Transformation of Variables.

229. A system of differential equations given in the symmetrical form is readily transformed so that a new variable replaces one of the given variables. For example, when there are three variables x, y, and z, let it be desired to replace x by a new variable u, a given function of x, y, and z. We have

$$\frac{dx}{X} = \frac{dy}{Y} = \frac{dz}{Z} = \frac{\lambda dx + \mu dy + \nu dz}{\lambda X + \mu Y + \nu Z}, \quad \ldots \ldots (1)$$

where λ, μ, and ν denote any arbitrary multipliers. Now, u being a given function of x, y, z,

$$du = \frac{du}{dx}dx + \frac{du}{dy}dy + \frac{du}{dz}dz.$$

Hence, if λ, μ, ν be taken equal to the partial derivatives of u, the numerator of the last fraction in equation (1) is du, and denoting the denominator by U, we have

$$\frac{dy}{Y} = \frac{dz}{Z} = \frac{du}{U}, \quad \dots \quad \dots \quad (2)$$

in which Y, Z, and U are to be expressed in terms of y, z, and u by the elimination of x.

As an illustration, in the example of the preceding article we may write

$$\frac{dx}{y} = \frac{dy}{x} = \frac{dz}{z} = \frac{dx + dy}{y + x};$$

so that, taking $u = x + y$, we have for one of the equations

$$\frac{dz}{z} = \frac{du}{u},$$

of which the integral is

$$u = bz,$$

which is equivalent to equation (3) of the preceding article.

Exact Equations.

230. If λ, μ, ν in equation (1), Art. 229, be so taken that

$$\lambda X + \mu Y + \nu Z = 0,$$

we shall have

$$\lambda dx + \mu dy + \nu dz = 0.$$

An equation derived in this manner may be exact, and thus lead directly to an integral equation containing all three of the variables.

For example, if the given equations are

$$\frac{dx}{mz - ny} = \frac{dy}{nx - lz} = \frac{dz}{ly - mx}, \quad \ldots \ldots (1)$$

we thus obtain

$$l\,dx + m\,dy + n\,dz = 0, \quad \ldots \ldots (2)$$

and also

$$x\,dx + y\,dy + z\,dz = 0. \quad \ldots \ldots (3)$$

Each of these is an exact equation, and their integration gives

$$lx + my + nz = a, \quad \ldots \ldots (4)$$

and

$$x^2 + y^2 + z^2 = b, \quad \ldots \ldots (5)$$

which constitute the complete solution of the given equations.

The Integrals of a System.

231. Denoting an exact equation derived as in the preceding article from the system

$$\frac{dx}{X} = \frac{dy}{Y} = \frac{dz}{Z} \quad \ldots \ldots (1)$$

by $du = 0$, the multipliers λ, μ, ν are the partial derivatives of the function u, and the relation connecting them is

$$X\frac{du}{dx} + Y\frac{du}{dy} + Z\frac{du}{dz} = 0. \quad \ldots \ldots (2)$$

Hence, if a function u satisfies this condition, the exact equation $du = 0$ is derivable from the system (1), and its integral

$$u = a$$

may be taken as one of the two equations which constitute the solution.

An equation of this form containing but one constant of integration is called an *integral* of the system in contradistinction from an integral equation which, like equation (3), Art. 228, contains more than one arbitrary constant.

Conversely, if $u = a$ is an integral of the system (1), the function u must satisfy equation (2): for let us transform the system as in Art. 229; then, because $du = 0$, we shall have $U = 0$, which is equation (2).

232. When there are more than three variables, we can derive in the same way a similar condition which must be satisfied by the partial derivatives of the function u, when $u = a$ is an integral. Thus it is possible to verify a single integral of a system without having a complete solution. The complete solution of a system involving $n + 1$ variables may be put in the form of a system of n integrals corresponding to the n arbitrary constants. The number of integrals is, however, in any case unlimited; for in the complete solution we may replace any constant by any function of the several constants. Thus, let

$$u = a \quad \text{and} \quad v = b$$

be two independent integrals of a system involving three variables, and let ϕ denote any function, then

$$\phi(u, v) = \phi(a, b) = C$$

is a relation between x, y, z and the arbitrary constant C, and is therefore an integral. This is, in fact, the general expression for the integrals of the system of which $u = a$ and $v = b$ are two independent integrals. Accordingly, it will be found that, if u and v are functions of x and y satisfying equation (2) of the preceding article, $\phi(u, v)$ also satisfies that equation, ϕ being an arbitrary function.

Equations of Higher Order equivalent to Determinate Systems of the First Order.

233. An equation of the second order may be regarded as equivalent to two equations of the first order between x, y and p, one of which is that which defines p, namely,

$$p = \frac{dy}{dx},$$

and the other is the result of writing $\frac{dp}{dx}$ in place of $\frac{d^2y}{dx^2}$ in the given equation. For example, the system equivalent to the equation

$$\frac{d^2y}{dx^2} + y = 0,$$

which is solved in Art. 76, is, when written in the symmetrical form of Art. 227,

$$\frac{dy}{p} = dx = -\frac{dp}{y},$$

in which the equation involving dp and dy is independent of x, and thus directly integrable.

The integrals of the equivalent system are the same as the *first integrals* of the equation of the second order, of which two, corresponding to the constants of integration employed, may be regarded as independent. Compare Art. 79. The complete integral of the equation of the second order, containing as it does both constants of integration, is an integral equation, but not an integral, being the result of eliminating the variable p either before or after a second integration. Compare Art. 82.

In like manner, an equation of the nth order is equivalent to a system of n equations of the first order, between $n + 1$ variables. Again, two simultaneous equations of the second order between three variables are equivalent to a system of four equations of the first order between five variables, and so on.

Geometrical Meaning of a System involving Three Variables.

234. Let x, y and z be regarded as the rectangular coordinates in space of a moving point; then, since the system of differential equations

$$\frac{dx}{X} = \frac{dy}{Y} = \frac{dz}{Z}$$

determines the ratios of dx, dy and dz, it determines at every instant the direction in which the point (x, y, z), subject to the differential equations, is moving. Starting, then, from any initial point A, the moving point will describe a definite line, and any two equations between x, y and z, representing two surfaces of which this line is the intersection, will form a particular solution. If we take a point not on the line thus determined for a new initial point, we shall determine another line in space representing another particular solution. The two equations forming the complete solution must contain two arbitrary constants, so that it may be possible to give any initial position to (x, y, z). The entire system of lines representing particular solutions is therefore a doubly infinite system of lines, no two of which can intersect, assuming X, Y and Z to be one-valued functions, because at each position there is but one direction in which the point (x, y, z) can move. We hence infer also that the constants will appear only in the first degree.

235. Consider, now, the complete solution as given by two integral equations between x, y, z and the constants a and b. The surfaces represented determine by their intersection a particular line of the system. Let the constant b pass through all possible values, while a remains fixed; then at least one of the surfaces moves, and the intersection describes a surface. The

equation of this surface is the integral corresponding to the constant a; for it is the result of eliminating b from the two equations, and is thus a relation between x, y, z and a. Hence, an integral represents a surface passing through a singly infinite system of lines selected from the doubly infinite system, and of course not intersecting any of the other lines of the system.*

If a and b both vary but in such a manner that $C = \phi(a, b)$ remains constant, the intersection of the two surfaces describes the surface whose equation is the integral corresponding to the constant C. Compare Art. 232.

236. Thus, in the example given in Art. 230, the integral (4) represents a plane perpendicular to the line

$$\frac{x}{l} = \frac{y}{m} = \frac{z}{n}, \quad \ldots \ldots \ldots (1)$$

and the integral (5) represents a sphere whose centre is at the origin. The intersection of the plane and sphere corresponding to particular values of the constants is a circle having its centre upon, and its plane perpendicular to, the fixed line (1).

Hence the doubly infinite system of lines represented by the differential equations (1), Art. 230, consists of the circles which have this line for axis; and the integrals of the differential system represent all surfaces of revolution having the same line for axis.

Examples XVIII.

Solve the following systems of simultaneous equations:—

1. $\dfrac{dx}{x} = \dfrac{dy}{z} = -\dfrac{dz}{y}$, $\qquad y^2 + z^2 = a$, $\log bx = \tan^{-1}\dfrac{y}{z}$.

---

* On the other hand, of the surface represented by an integral equation, we can only say that it passes through a particular line of the system.

EXAMPLES.

2. $\begin{cases} \dfrac{dx}{dt} + \dfrac{2x}{t} = 1, \\ \dfrac{dy}{dt} = x + y + \dfrac{2x}{t} - 1, \end{cases}$ \qquad $x = \dfrac{t}{3} + \dfrac{a}{t^2},$
$x + y = be^t.$

3. $\dfrac{dx}{y+z} = \dfrac{dy}{z+x} = \dfrac{dz}{x+y},$ \qquad $\sqrt{(x+y+z)} = \dfrac{a}{z-y} = \dfrac{b}{x-z}.$

4. $\dfrac{dx}{x^2 - y^2 - z^2} = \dfrac{dy}{2xy} = \dfrac{dz}{2xz},$ \qquad $y = az, \quad x^2 + y^2 + z^2 = bz.$

5. $\dfrac{ldx}{mn(y-z)} = \dfrac{mdy}{nl(z-x)} = \dfrac{ndz}{lm(x-y)},$ \qquad $l^2x + m^2y + n^2z = a,$
$l^2x^2 + m^2y^2 + n^2z^2 = b.$

6. $\dfrac{adx}{(b-c)yz} = \dfrac{bdy}{(c-a)zx} = \dfrac{cdz}{(a-b)xy},$ \qquad $ax^2 + by^2 + cz^2 = A,$
$a^2x^2 + b^2y^2 + c^2z^2 = B.$

7. $\dfrac{dx}{x} = \dfrac{dy}{y} = \dfrac{dz}{z - a\sqrt{(x^2+y^2+z^2)}},$

$y = ax, \quad x^{1-a} = \beta[z + \sqrt{(x^2+y^2+z^2)}].$

8. Show that the general integral of
$$\dfrac{dx}{l} = \dfrac{dy}{m} = \dfrac{dz}{n}$$
represents cylindrical surfaces, and that the general integral of
$$\dfrac{dx}{x-a} = \dfrac{dy}{y-\beta} = \dfrac{dz}{z-\gamma}$$
represents conical surfaces.

XIX.

Simultaneous Linear Equations.

237. We have seen that the complete solution of a system of simultaneous equations of the first order between $n+1$ variables consists of n relations between the $n+1$ variables and n constants of integration. Selecting any two variables, the elimination of the remaining $n-1$ variables gives a relation between these two variables, involving in general the n constants.

We may also, selecting one of the two variables as independent, perform the elimination before the integration, the result being the equation of the nth order,* of which the equation just mentioned is the complete integral.

For example, in the case of three variables, x, y and t, if we require the differential equation connecting x with the independent variable t, the two given equations are to be regarded as connecting with t the four quantities x, y, $\dfrac{dx}{dt}$ and $\dfrac{dy}{dt}$. Taking their derivatives with respect to t, we have four equations containing x, y, $\dfrac{dx}{dt}$, $\dfrac{dy}{dt}$, $\dfrac{d^2x}{dt^2}$ and $\dfrac{d^2y}{dt^2}$; and from these four we can eliminate y, $\dfrac{dy}{dt}$ and $\dfrac{dy^2}{dt^2}$, thus obtaining an equation of the second order, in which x is the dependent, and t the independent variable.

238. As a method of solution the process is particularly applicable to linear equations with constant coefficients, since

---

* The differential equation connecting two of the variables may be of a lower order, in which case the integral relation will contain fewer than n constants. For example, one of the equations of the first order may contain only two variables, as in Art. 228, and then the integral relation will contain but one constant.

in that case we have a direct method of solving the resulting equations.

For example, the equations

$$\frac{dx}{dt} + 5x + y = e^t \quad \ldots \ldots \ldots (1)$$

and

$$\frac{dy}{dt} - x + 3y = e^{2t} \quad \ldots \ldots \ldots (2)$$

are linear equations with constant coefficients, if t be taken as the independent variable. Differentiating the first equation, we have

$$\frac{d^2x}{dt^2} + 5\frac{dx}{dt} + \frac{dy}{dt} = e^t, \quad \ldots \ldots (3)$$

and since $\frac{d^2y}{dt^2}$ does not occur in this it is unnecessary to differentiate the second. Eliminating $\frac{dy}{dt}$ and y by means of equations (2) and (1), we have

$$\frac{d^2x}{dt^2} + 8\frac{dx}{dt} + 16x = 4e^t - e^{2t}.$$

The complementary function is $(A + Bt)e^{-4t}$, and the particular integral is found by the methods of section X. The resulting value of x is

$$x = (A + Bt)e^{-4t} + \tfrac{4}{25}e^t - \tfrac{1}{36}e^{2t},$$

and, substituting this value in equation (1), we find without further integration,

$$y = -(A + B + Bt)e^{-4t} + \tfrac{7}{36}e^{2t} + \tfrac{1}{25}e^t.$$

239. The differentiation and elimination required in the process illustrated above are more expeditiously performed by the symbolic method. For, since the differentiation is indicated by symbolic multiplication by D, the equations may be treated as ordinary algebraic equations. Moreover, the process is the same if one or both the equations are of an order higher than the first.

For example, the system

$$2\frac{d^2y}{dt^2} - \frac{dx}{dt} - 4y = 2t,$$

$$4\frac{dx}{dt} + 2\frac{dy}{dt} - 3x = 0,$$

when written symbolically, is

$$(2D^2 - 4)y - Dx = 2t,$$

$$2Dy + (4D - 3)x = 0.$$

Eliminating x, we have, in the determinant notation,

$$\begin{vmatrix} 2D^2 - 4 & -D \\ 2D & 4D - 3 \end{vmatrix} y = \begin{vmatrix} 2t & -D \\ 0 & 4D - 3 \end{vmatrix},$$

or

$$(D - 1)^2 (2D + 3)y = 2 - \tfrac{3}{2}t,$$

and integrating,

$$y = (A + Bt)e^t + Ce^{-\frac{3}{2}t} - \tfrac{1}{2}t.$$

The value of x is, in this example, most readily derived from that of y by first eliminating Dx from the given equations, thus obtaining

$$(8D^2 + 2D - 16)y - 3x = 8t,$$

whence, substituting the value of y,

$$x = e^t(6B - 2A - 2Bt) - \tfrac{1}{3}Ce^{-\frac{3}{2}t} - \tfrac{1}{3}.$$

240. Ordinarily, in finding the value of the variable first eliminated it is necessary to perform an integration, and, when this is done, the new constants of integration are not arbitrary, but must be determined so as to satisfy the given equations. Thus, if in the preceding example the value of x had been derived from the first of the given equations, after substituting the value of y, it would have contained an unknown constant in place of the term $-\frac{1}{8}$, and it would have been necessary to substitute in the second equation to determine the value of this constant.

The value of x may also be derived directly from the result of eliminating y, namely,

$$\begin{vmatrix} 2D^2 - 4 & -D \\ 2D & 4D - 3 \end{vmatrix} x = \begin{vmatrix} 2D^2 - 4 & 2t \\ 2D & 0 \end{vmatrix}.$$

The complementary functions for the two variables will then be of the same form, and will involve two sets of constants. By substituting in one of the given equations, we shall have an identity in which, equating to zero the coefficients of the several terms of the complementary function, the relations between the constants may be determined.

241. The number of constants of integration which enter the solution is that which indicates the order of the resultant equation. This number is not necessarily the sum of the indices of the orders of the given equations, although it cannot exceed this sum; it depends upon the form of the given equations, being, as the process shows, the index of the degree in D of the determinant of the first members.

Denoting this number by m, the values of the n dependent variables contain n sets of m constants, of which one set is arbitrary. Substituting the values in one of the given equations, we have an identity giving m relations between the constants; it is therefore necessary to substitute in $n - 1$ of the given equations to obtain the relations between the constants.

Introduction of a New Variable.

242. The solution of a system of differential equations is sometimes facilitated by the introduction of a new variable, in terms of which we then seek to express each of the original variables. Given, for example, the system

$$\frac{dx}{X} = \frac{dy}{Y} = \frac{dz}{Z}, \quad \ldots \ldots \ldots (1)$$

where

$$X = ax + by + cz + d, \qquad Y = a'x + b'y + c'z + d',$$

$$Z = a''x + b''y + c''z + d''.$$

If we introduce a new variable t by assuming dt equal to the common value of the members of equation (1), we shall have the system

$$\frac{dx}{X} = \frac{dy}{Y} = \frac{dz}{Z} = dt, \quad \ldots \ldots \ldots (2)$$

involving four variables, which is linear if t be taken as the independent variable. Writing the equations symbolically, the system is

$$\left. \begin{array}{l} (a-D)x + by + cz + d = 0, \\ a'x + (b'-D)y + c'z + d' = 0, \\ a''x + b''y + (c''-D)z + d'' = 0; \end{array} \right\} \ldots \ldots (3)$$

whence

$$\begin{vmatrix} a-D & b & c \\ a' & b'-D & c' \\ a'' & b'' & c''-D \end{vmatrix} x = - \begin{vmatrix} d & b & c \\ d' & b'-D & c' \\ d'' & b'' & c''-D \end{vmatrix}, \ldots (4)$$

§ XIX.]   INTRODUCTION OF A NEW VARIABLE.   265

in which D may be omitted in the second member because it contains no variable. Denoting the roots of the cubic

$$\begin{vmatrix} a-D & b & c \\ a' & b'-D & c' \\ a'' & b'' & c''-D \end{vmatrix} = 0 \quad \ldots \quad (5)$$

by λ_1, λ_2 and λ_3, equation (4) and the similar equations for y and z give

$$\left. \begin{array}{l} x = Ae^{\lambda_1 t} + Be^{\lambda_2 t} + Ce^{\lambda_3 t} + k \\ y = A'e^{\lambda_1 t} + B'e^{\lambda_2 t} + C'e^{\lambda_3 t} + k' \\ z = A''e^{\lambda_1 t} + B''e^{\lambda_2 t} + C''e^{\lambda_3 t} + k'' \end{array} \right\}, \quad \ldots \quad (6)$$

in which k, k', k'' are the values of x, y, z respectively, which make $X = 0$, $Y = 0$ and $Z = 0$.

Substituting these values in the first of equations (3), we have one of the three equations determining k, k' and k'', and for the constants of integration the three relations,

$$(a - \lambda_1)A + bA' + cA'' = 0,$$
$$(a - \lambda_2)B + bB' + cB'' = 0,$$
$$(a - \lambda_3)C + bC' + cC'' = 0.$$

In like manner, substitution in each of the other equations gives three relations between the constants, making in all nine relations, of which six are independent. The three relations between A, A' and A'' are

$$(a - \lambda_1)A + bA' + cA'' = 0,$$
$$a'A + (b' - \lambda_1)A' + c'A'' = 0,$$
$$a''A + b''A' + (c'' - \lambda_1)A'' = 0,$$

which are equivalent to two equations for the ratios $A : A' : A''$, since their determinant vanishes because λ_1 is a root of equation (5).

243. The introduction of a new variable, as in the preceding article, introduces a new constant of integration into the system, but this constant is so connected with the new variable that the relations between the original variables obtained by eliminating the new variable are independent also of this constant. Thus in the value of x, equation (6), we might have put $t + a$ in place of t, employing only two other constants; then the relations between x, y and z, which we should obtain by eliminating t, would obviously contain only the two constants last mentioned.

EXAMPLES XIX.

Solve the following systems of linear equations:—

1. $\dfrac{dx}{dt} + 5x - 2y = e^t$, $\dfrac{dy}{dt} - x + 6y = e^{2t}$,

$$x = Ae^{-4t} + Be^{-7t} + \tfrac{7}{40}e^t + \tfrac{1}{27}e^{2t},$$
$$y = \tfrac{1}{2}Ae^{-4t} - Be^{-7t} + \tfrac{1}{40}e^t + \tfrac{7}{54}e^{2t}.$$

2. $\dfrac{dx}{3x - y} = \dfrac{dy}{x + y} = dt$,

$$x = (A + Bt)e^{2t}, \quad y = (A - B + Bt)e^{2t}.$$

3. $(5y + 9z)dx + dy + dz = 0$, $(4y + 3z)dx + 2dy - dz = 0$,

$$y = Ae^{-x} + Be^{-7x}, \quad z = -\tfrac{1}{2}Ae^{-x} + Be^{-7x}.$$

4. $\dfrac{dx}{-my} = dt = \dfrac{dy}{mx}$,

$$x = A\cos mt + B\sin mt, \quad y = A\sin mt - B\cos mt.$$

5. $a\dfrac{dz}{dx} + n^2y = e^x,\quad \dfrac{dy}{dx} + az = 0,$

$$y = Ae^{nx} + Be^{-nx} + \dfrac{e^x}{n^2 - 1},$$

$$az = -nAe^{nx} + nBe^{-nx} - \dfrac{e^x}{n^2 - 1}.$$

6. $\dfrac{d^2x}{dt^2} + m^2y = 0,\quad \dfrac{d^2y}{dt^2} - m^2x = 0,$

$$x = e^{\frac{mx}{\sqrt{2}}}\left(A_1 \cos\dfrac{mx}{\sqrt{2}} + A_2 \sin\dfrac{mx}{\sqrt{2}}\right) + e^{-\frac{mx}{\sqrt{2}}}\left(A_3 \cos\dfrac{mx}{\sqrt{2}} + A_4 \sin\dfrac{mx}{\sqrt{2}}\right),$$

$$y = e^{\frac{mx}{\sqrt{2}}}\left(A_1 \sin\dfrac{mx}{\sqrt{2}} - A_2 \cos\dfrac{mx}{\sqrt{2}}\right) + e^{-\frac{mx}{\sqrt{2}}}\left(A_4 \cos\dfrac{mx}{\sqrt{2}} - A_3 \sin\dfrac{mx}{\sqrt{2}}\right).$$

7. $\begin{cases} 4\dfrac{dx}{dt} + 9\dfrac{dy}{dt} + 44x + 49y = t, \\[4pt] 3\dfrac{dx}{dt} + 7\dfrac{dy}{dt} + 34x + 38y = e^t, \end{cases}$

$$x = Ae^{-t} + Be^{-6t} + \tfrac{19}{3}t - \tfrac{56}{9} - \tfrac{29}{7}e^t,$$

$$y = -Ae^{-t} + 4Be^{-6t} - \tfrac{17}{3}t + \tfrac{55}{9} + \tfrac{24}{7}e^t.$$

8. $\dfrac{dx}{dt} + \dfrac{dy}{dt} + 2x + y = 0,\quad \dfrac{dy}{dt} + 5x + 3y = 0,$

$$y = A\cos t + B\sin t,\quad x = -\dfrac{3A+B}{5}\cos t + \dfrac{A-3B}{5}\sin t.$$

9. $\dfrac{d^2x}{dt^2} - 2n\dfrac{dy}{dt} + m^2x = 0,\quad \dfrac{d^2y}{dt^2} + 2n\dfrac{dx}{dt} + m^2y = 0,$

$$x = A_1 \cos at - A_2 \sin at + B_1 \cos \beta t - B_2 \sin \beta t,$$

$$y = A_2 \cos at + A_1 \sin at + B_2 \cos \beta t + B_1 \sin \beta t;$$

where a and β stand for $-n \pm \sqrt{(n^2 + m^2)}$.

10. $\begin{cases} \dfrac{dx_1}{dt} = a_3 x_3 - a_1 x_1, \\ \dfrac{dx_2}{dt} = a_1 x_1 - a_2 x_2, \\ \dfrac{dx_3}{dt} = a_2 x_2 - a_3 x_3, \end{cases}$

$$x_1 = Ae^{\lambda_1 t} + Be^{\lambda_2 t} + C,$$

$$x_2 = \frac{a_1 A}{\lambda_1 + a_2} e^{\lambda_1 t} + \frac{a_1 B}{\lambda_2 + a_2} e^{\lambda_2 t} + \frac{a_1 C}{a_2},$$

$$x_3 = \frac{a_2 a_1 A}{(\lambda_1 + a_3)(\lambda_1 + a_2)} e^{\lambda_1 t} + \frac{a_2 a_1 B}{(\lambda_2 + a_3)(\lambda_2 + a_2)} e^{\lambda_2 t} + \frac{a_1 C}{a_3},$$

where λ_1 and λ_2 are the roots of

$$\lambda^2 + (a_1 + a_2 + a_3)\lambda + a_1 a_2 + a_2 a_3 + a_3 a_1 = 0.$$

11. $\dfrac{dx}{dt} = 7x - y, \quad \dfrac{dy}{dt} = 2x + 5y,$

$$x = e^{6t}(A \cos t + B \sin t),$$
$$y = e^{6t}[(A - B)\cos t + (A + B)\sin t].$$

12. $\begin{cases} \dfrac{dx}{dt} + 2\dfrac{dy}{dt} + x + 7y = e^t - 3, \\ \dfrac{dy}{dt} - 2x + 3y = 12 - 3e^t, \end{cases}$

$$x = Ae^{-4t}\cos t + Be^{-4t}\sin t + \tfrac{81}{26}e^t - \tfrac{99}{17},$$
$$y = -(A+B)e^{-4t}\cos t + (A-B)e^{-4t}\sin t - \tfrac{2}{13}e^t + \tfrac{6}{17}.$$

13. $t\dfrac{dx}{dt} + 2x - 2y = t, \quad t\dfrac{dy}{dt} + x + 5y = t^2,$

$$x = At^{-4} + Bt^{-3} + \tfrac{3}{10}t + \tfrac{1}{15}t^2,$$
$$y = -At^{-4} - \tfrac{1}{2}Bt^{-3} - \tfrac{1}{20}t + \tfrac{2}{15}t^2.$$

14. $\dfrac{d^2x}{dt^2} - 3x - 4y = 0$, $\dfrac{d^2y}{dt^2} + x + y = 0$,

$$x = (A_1 + B_1 t)e^t + (A_2 + B_2 t)e^{-t},$$
$$2y = (B_1 - A_1 - B_1 t)e^t - (A_2 + B_2 + B_2 t)e^{-t}.$$

15. $\dfrac{d^2x}{dt^2} - \dfrac{dy}{dt} - 8x = 8t$, $\dfrac{dx}{dt} + 2\dfrac{dy}{dt} - 3y = 0$,

$$x = (A + Bt)e^{2t} + Ce^{-3t} - t,$$
$$y = (3B - 2A - 2Bt)e^{2t} - \tfrac{1}{8}Ce^{-3t} - \tfrac{1}{8}.$$

16. Show that the integrals of the system

$$\dfrac{dx}{dt} = ax + by + c, \qquad \dfrac{dy}{dt} = a'x + b'y + c',$$

are

$$(a + m_1 a')(x + m_1 y) + c + m_1 c' = A_1 e^{(a + m_1 a')t},$$
$$(a + m_2 a')(x + m_2 y) + c + m_2 c' = A_2 e^{(a + m_2 a')t},$$

where m_1 and m_2 are the roots of

$$a' m^2 + (a - b') m - b = 0;$$

and obtain a similar solution for the system

$$\dfrac{d^2x}{dt^2} = ax + by, \qquad \dfrac{d^2y}{dt^2} = a'x + b'y,$$

$$x + m_1 y = A_1 e^{(a + m_1 a')^{\frac{1}{2}} t} + B_1 e^{-(a + m_1 a')^{\frac{1}{2}} t},$$
$$x + m_2 y = A_2 e^{(a + m_2 a')^{\frac{1}{2}} t} + B_2 e^{-(a + m_2 a')^{\frac{1}{2}} t}.$$

XX.

Single Differential Equations involving more than Two Variables.

244. When the number of differential equations connecting $n + 1$ variables is less than n, it is of course impossible to establish n integral relations between the variables. We shall here consider only the case of a single equation, at first supposing the number of variables to be three; and we shall find that there does not always exist an equivalent single integral relation between the variables.

We have seen that when there are two differential relations between x, y and z, the integrable equations which separately furnish the two independent relations between the variables are generally produced by the combination of the given equations. We have now to find the condition under which a single given equation is thus integrable, and the meaning of an equation in which the condition is not fulfilled.

The Condition of Integrability.

245. The given equation will be of the form

$$Pdx + Qdy + Rdz = 0, \quad \ldots \ldots \quad (1)$$

in which P, Q and R may be any functions of x, y and z. If there be an integral relation between x, y, z and an arbitrary constant a to which this equation is equivalent, let it be put in the form

$$u = a,$$

so that a shall disappear by differentiation; then the differential equation $du = 0$, or

$$\frac{du}{dx}dx + \frac{du}{dy}dy + \frac{du}{dz}dz = 0,$$

§ XX.] THE CONDITION OF INTEGRABILITY. 271

must be equivalent to equation (1). In other words, if the equation is integrable, there must exist a function of x, y and z whose partial derivatives are proportional to P, Q and R; thus

$$\frac{du}{dx} = \mu P, \quad \frac{du}{dy} = \mu Q, \quad \frac{du}{dz} = \mu R.$$

Now, since $\dfrac{d}{dy}\dfrac{du}{dx} = \dfrac{d}{dx}\dfrac{du}{dy}$, etc., these equations give

$$\mu\left(\frac{dP}{dy} - \frac{dQ}{dx}\right) = Q\frac{d\mu}{dx} - P\frac{d\mu}{dy},$$

$$\mu\left(\frac{dQ}{dz} - \frac{dR}{dy}\right) = R\frac{d\mu}{dy} - Q\frac{d\mu}{dz},$$

$$\mu\left(\frac{dR}{dx} - \frac{dP}{dz}\right) = P\frac{d\mu}{dz} - R\frac{d\mu}{dx}.$$

Multiplying the first of these equations by R, the second by P and the third by Q, and adding the results, μ is eliminated, and we have

$$P\left(\frac{dQ}{dz} - \frac{dR}{dy}\right) + Q\left(\frac{dR}{dx} - \frac{dP}{dz}\right) + R\left(\frac{dP}{dy} - \frac{dQ}{dx}\right) = 0, \quad . \quad (2)$$

for the condition under which the equation (1) admits of an integral.*

---

* If the given equation is exact, the three equations above are satisfied by $\mu = 1$, and each of the binomials in equation (2) vanishes. If one of the binomials vanishes while equation (2) is satisfied, an integrating factor which is a function of one variable only exists, and in this case μ is readily determined.

In general, if μ is an integrating factor and $u = a$ is the corresponding integral, $F(u) = F(a)$, where F is any function, is also an integral, and $\mu F'(u)$ is the corresponding integrating factor. Thus $\mu f(u)$ is the general expression for the integrating factor.

Solution of the Integrable Equation.

246. Supposing the condition of integrability to be satisfied, if in the integral one of the variables, say z, be regarded as constant, the corresponding differential equation between x and y will be

$$Pdx + Qdy = 0.$$

Hence the complete integral of this equation will include the integral sought if the constant of integration be regarded as a function of z. Finally, this function of z may be determined by comparing the total differential equation of the complete integral with the given equation.

Given, for example, the equation

$$zydx - zxdy - y^2dz = 0, \quad \ldots \ldots (1)$$

in which $P = zy$, $Q = -zx$, $R = -y^2$, and the condition of integrability is found to be satisfied. Treating z as a constant, the equation becomes

$$ydx - xdy = 0, \quad \ldots \ldots \ldots (2)$$

of which the complete integral is

$$x - Cy = 0. \quad \ldots \ldots \ldots (3)$$

This is, therefore, the integral of equation (1), C being independent of x and y but involving z. Differentiating, we have

$$dx - Cdy - ydC = 0.$$

Multiplying by zy to make the term containing dx identical with that in equation (1), the coefficients of dy are identical by virtue of equation (3), and the equations agree if $-zy^2dC = -y^2dz$, or

$$dC = \frac{dz}{z}.$$

§ XX.]   SOLUTION OF THE INTEGRABLE EQUATION.   273

Hence $C = c + \log z$, and the integral (3) becomes

$$x - cy - y \log z = 0,$$

which is the integral of equation (1).

247. It is to be noticed that the possibility of obtaining a differential relation between C and z, independent of x and y, sufficiently indicates the integrability of the equation. But in some cases it is necessary, in order to obtain such an equation, to eliminate the other variables from the equation containing dC by means of the integral itself. Thus, let the given equation be

$$x dx + z dz = \sqrt{(h^2 - x^2 - z^2)} dy. \quad \ldots \quad (1)$$

If y be regarded as constant, the integral is

$$x^2 + z^2 = C. \quad \ldots \ldots \ldots (2)$$

This is therefore the integral of equation (1), if it be possible to determine C as a function of y. Differentiating and comparing with the given equation, we find

$$\tfrac{1}{2} dC = \sqrt{(h^2 - x^2 - z^2)} dy,$$

an equation containing x and z; but, eliminating x by means of equation (2), z also disappears, and we have

$$\frac{dC}{2\sqrt{(h^2 - C)}} = dy.$$

Hence

$$\sqrt{(h^2 - C)} = -y + c,$$

and, substituting the value of C thus determined in equation (2), we have, for the integral of equation (1),

$$x^2 + z^2 + (y - c)^2 = h^2.$$

Separation of the Variables.

248. When it is possible to put the equation in such a form that one of the variables occurs only in an exact differential, the equation will, *if integrable*, be thus rendered exact. Suppose, for example, that it can be written in the form

$$dw + S\,dx + T\,dy = 0, \quad \ldots \ldots \quad (1)$$

in which S and T are independent of z. Now, if there be an integral, it may be put in the form $z = f(x, y)$; hence, also, by substituting this value of z in the expression for w as a function of x, y and z, it may be put in the form

$$w + \phi(x, y) = 0. \quad \ldots \ldots \quad (2)$$

The differential of this equation must be identical with equation (1), because the terms containing dz are identical; therefore, if equation (1) be integrable, it is already exact, and its integral is

$$w + \phi(x, y) = c.$$

In fact, the condition of integrability, Art. 245, reduces in this case to

$$\frac{dS}{dy} - \frac{dT}{dx} = 0,$$

which is the same as the condition of exactness for the differential expression $S\,dx + T\,dy$. See Art. 25.

249. The most obvious application of this principle is to the case in which one variable can be entirely separated from the other two. Thus the example in Art. 246 might have been solved in this way; for, dividing by zy^2, which separates the variable z, it becomes

$$\frac{y\,dx - x\,dy}{y^2} - \frac{dz}{z} = 0,$$

an exact equation of which the integral is

$$\frac{x}{y} - \log z = c.$$

Homogeneous Equations.

250. In the case of a homogeneous equation between x, y and z, one variable can be separated from the other two by means of a transformation of the same form as that employed in the corresponding case with two variables, Art. 20. For, putting

$$x = zu, \qquad y = zv,$$

the homogeneous equation may be written in the form

$$z^n \phi(u, v) dx + z^n \psi(u, v) dy + z^n \chi(u, v) dz = 0;$$

and, substituting

$$dx = z du + u dz, \qquad dy = z dv + v dz,$$

we have

$$z\phi(u, v) du + z\psi(u, v) dv + [\chi(u, v) + u\phi(u, v) + v\psi(u, v)] dz = 0.$$

If the coefficient of dz vanishes, we have an equation between the two variables u and v. If not, the equation takes the form

$$\frac{dz}{z} + \frac{\phi(u, v) du + \psi(u, v) dv}{\chi(u, v) + u\phi(u, v) + v\psi(u, v)} = 0,$$

and, in accordance with Art. 248, the second term will be an exact differential if the given equation is integrable.

251. As an example, let us take the equation

$$(y^2 + yz + z^2) dx + (z^2 + zx + x^2) dy + (x^2 + xy + y^2) dz = 0, \quad (1)$$

which will be found to satisfy the condition of integrability. Making the substitutions, and reducing, we have

$$\frac{dz}{z} + \frac{(v^2 + v + 1)du + (u^2 + u + 1)dv}{(u + v + 1)(uv + u + v)} = 0.$$

Knowing the second term to be an exact differential, we integrate it at once with respect to u, and obtain

$$\log z - \log \frac{u + v + 1}{uv + u + v} + C = 0,$$

The symmetry of this equation shows that C is a constant and not a function of v: thus the integral of equation (1) is

$$xy + yz + zx = c(x + y + z).$$

Equations containing more than Three Variables.

252. In order that an equation of the form

$$Pdx + Qdy + Rdz + Tdt = 0$$

involving four variables may be integrable, it must obviously be integrable when any one of the four variables is made constant. Thus, regarding z, x and y successively as constants, equation (2), Art. 245, gives the three conditions of integrability,

$$T\left(\frac{dP}{dy} - \frac{dQ}{dx}\right) + P\left(\frac{dQ}{dt} - \frac{dT}{dy}\right) + Q\left(\frac{dT}{dx} - \frac{dP}{dt}\right) = 0,$$

$$T\left(\frac{dQ}{dz} - \frac{dR}{dy}\right) + Q\left(\frac{dR}{dt} - \frac{dT}{dz}\right) + R\left(\frac{dT}{dy} - \frac{dQ}{dt}\right) = 0,$$

$$T\left(\frac{dR}{dx} - \frac{dP}{dz}\right) + R\left(\frac{dP}{dt} - \frac{dT}{dx}\right) + P\left(\frac{dT}{dz} - \frac{dR}{dt}\right) = 0.$$

§ XX.] *MORE THAN THREE VARIABLES.* 277

Again regarding t as constant, we have the condition

$$P\left(\frac{dQ}{dz} - \frac{dR}{dy}\right) + Q\left(\frac{dR}{dx} - \frac{dP}{dz}\right) + R\left(\frac{dP}{dy} - \frac{dQ}{dx}\right) = 0;$$

but this is not an independent condition, for it may be deduced by multiplying the preceding equations by R, P and Q respectively, and adding the results.

253. In general, if the equation contains n variables, the number of conditions of the above form which we can write is $\frac{n(n-1)(n-2)}{1.2.3}$, which is the number of ways we can select three out of the n variables. But, in writing the independent conditions, we may confine our attention to those in which a selected variable occurs, for any condition not containing this variable may be obtained exactly as in the preceding article from three of those which do contain it. Thus the number of independent conditions is $\frac{(n-1)(n-2)}{1.2}$, which is the number of ways we can select two out of the $n-1$ remaining variables.

254. When the conditions of integrability are satisfied, the integral is found, as in the case of three variables, by first integrating as if all the variables except two were constant, the quantity C introduced by this integration being a function of those variables which were taken as constants. To determine this function the total differential of the result is compared with the given equation. The result either determines the value of dC in terms of these last variables (in which case dC should be an exact differential), or else is such that the first two variables may be eliminated simultaneously, as in the example of Art. 247, giving an integrable equation between C and the remaining variables.

The Non-Integrable Equation.

255. In an equation of the form

$$Pdx + Qdy + Rdz = 0$$

the variables x, y and z may have any simultaneous values whatever; but, for each set of values, the equation imposes a restriction upon the relative rates of variation of the variables, that is, upon the ratios of dx, dy and dz. When the condition expressed by equation (2), Art. 245, is satisfied, there exists an integral equation which, for each of the sets of values of x, y and z which satisfy it, imposes the same restriction upon their relative rates of variation. At the same time the presence of an arbitrary constant makes the integral sufficiently general to be satisfied by any simultaneous values of x, y and z.

But, when the condition of integrability is not satisfied, there is no such integral equation. Two integral equations will, however, constitute a particular solution, when, for each set of simultaneous values of x, y and z which satisfy them, the ratios which they determine for dx, dy and dz satisfy, in connection with these values, the given differential equation.

256. If one of the two integral equations is assumed in advance, the determination of the particular solutions consistent with the assumed equation is effected by solving a pair of simultaneous differential equations, namely, the given equation and the result of differentiating the assumed relation. Geometrically the problem is that of determining the lines upon a certain surface which satisfy the given differential equation.

For example, given the equation

$$(1 + 2a)xdx + y(1 - x)dy + zdz = 0 \quad \ldots \quad (1)$$

(which it will be found does not satisfy the condition of integrability); let it be required to find the lines on the surface of the sphere

$$x^2 + y^2 + z^2 = b^2 \quad \ldots \ldots \ldots (2)$$

such that a point moving along any one of them satisfies equation (1). Differentiating equation (2), we have

$$xdx + ydy + zdz = 0, \quad \ldots \ldots \ldots (3)$$

which with equation (1) forms a system of which equation (2) is one integral and a second integral is required. Subtracting, we have an equation free from z, namely,

$$2axdx - xydy = 0,$$

the integral of which is

$$y^2 = 4ax + C. \quad \ldots \ldots \ldots (4)$$

Hence the required lines are those whose projections upon the plane of xy are the parabolas represented by equation (4).

257. In order to form a general solution of a non-integrable equation, the assumed equation must contain an arbitrary function. We might, for example, assume

$$y = f(x), \quad \ldots \ldots \ldots (1)$$

where f is arbitrary, because any particular solution consisting of two relations between x, y and z might be put in the form $y = f(x)$, $z = \phi(x)$. If, therefore, we determine all the particular solutions consistent with equation (1), the result will, when f is regarded as arbitrary, include all the particular solutions. The equation which completes the solution will, as in the preceding example, be found by integration, and will therefore contain an arbitrary constant C, to which a special value must be given (as well as a special form to the function f) in order to produce a given particular solution.

258. The general solution of the equation

$$Pdx + Qdy + Rdz = 0 \quad \ldots \ldots (1)$$

may be presented in quite a different form, which is due to Monge, depending upon a special mode of assuming the equation containing the arbitrary function.

Let μ be an integrating factor of the equation

$$Pdx + Qdy = 0$$

when z is regarded as a constant, and let $V = C$ be the corresponding integral, so that

$$dV = \mu Pdx + \mu Qdy.$$

Then, in the first place, the pair of equations

$$z = c, \quad \text{and} \quad V = C, \ \ldots \ldots (2)$$

where c and C are arbitrary constants, constitutes a class of particular solutions of (1). Now, for the general solution, let us assume

$$V = \phi(z). \ \ldots \ldots \ldots (3)$$

Differentiating, we have

$$\mu Pdx + \mu Qdy + \left[\frac{dV}{dz} - \phi'(z)\right]dz = 0, \ \ldots \ldots (4)$$

which, combined with equation (1), gives

$$\left(\frac{dV}{dz} - \phi'(z) - \mu R\right)dz = 0. \ \ldots \ldots (5)$$

Hence, if $V = \phi(z)$ be taken as one of the relations between the variables, we must have, in order to satisfy equation (1), either $dz = 0$, or else

$$\frac{dV}{dz} - \phi'(z) - \mu R = 0. \quad \ldots \ldots \ldots (6)$$

The first supposition gives $z = c$ and $V = \phi(c)$, a system of solutions of the form (2); the second constitutes, in connection with equation (3), *Monge's solution*.

It is to be noticed that when it is possible to determine ϕ so that equation (6) is *identically* satisfied, the given equation is integrable, and $V = \phi(z)$ is its integral. But, in the non-integrable case, ϕ is to be regarded as arbitrary.

Monge's solution includes all solutions excepting those of the form (2). To show this, it is only necessary to notice that, with this exception, any particular solution can be expressed in the form $x = f_1(z)$, $y = f_2(z)$; and, substituting these values in the expression for V as a function of x, y and z, we have an equation of the form $V = \phi(z)$ determining the form of ϕ for the particular solution in question. The particular solution is therefore among those determined by one of the two methods of satisfying equation (5); and, as it is not of the form (2), it must be that determined by equations (3) and (6).

The distinction between this solution and that given in Art. 257 is further explained in Art. 262 from the geometrical point of view.

Geometrical Meaning of a Single Differential Equation between Three Variables.

259. Regarding x, y and z as the rectangular coordinates of a variable point, as in Art. 234, the single equation

$$Pdx + Qdy + Rdz = 0 \quad \ldots \ldots \ldots (1)$$

expresses that the point (x, y, z) is moving in some direction, of which the direction-cosines l, m, n, which are proportional to dx, dy and dz, satisfy the condition

$$Pl + Qm + Rn = 0. \quad \ldots \ldots \quad (2)$$

Consider also a point satisfying the simultaneous equations

$$\frac{dx}{P} = \frac{dy}{Q} = \frac{dz}{R}, \quad \ldots \ldots \quad (3)$$

and therefore moving in the direction whose direction-cosines satisfy

$$\frac{\lambda}{P} = \frac{\mu}{Q} = \frac{\nu}{R}. \quad \ldots \ldots \quad (4)$$

Suppose the moving points which satisfy equations (1) and (3) respectively to be passing through the same fixed point A; then P, Q and R have the same values for each, and equations (2) and (4) give

$$l\lambda + m\mu + n\nu = 0,$$

which is the condition expressing that the directions in question are at right angles. We have seen, in Art. 234, that equations (3) represent a system of lines, there being one line of the system passing through any given point. Hence equation (1) simply restricts a point to move in such a manner that it everywhere cuts orthogonally the system of lines represented by equations (3), which we may call the *auxiliary system*.

260. Now, suppose in the first place that equation (1) is integrable. The integral represents a system of surfaces one of which passes through the given point A. This surface contains all the possible paths of the moving point which pass through A, and every line in space representing a particular solution lies in some one of the surfaces belonging to the system.

§ XX.] *GEOMETRICAL INTERPRETATION.* 283

The restriction imposed by equation (1) is in this case completely expressed by a single equation.

Every member of the system of surfaces represented by the integral cuts the auxiliary system of lines orthogonally, so that equation (2), Art. 245, considered with reference to the system of lines represented by equations (3), expresses the condition that the system shall admit of a system of orthogonally cutting surfaces.

261. On the other hand, when the condition of integrability is not satisfied, the possible paths of the moving point which pass through A do not lie in any one surface, the auxiliary system of lines, in this case, not admitting of orthogonally cutting surfaces.*

When, as in the example of Art. 256, the point subject to equation (1) is in addition restricted to a given surface, the auxiliary lines not piercing this surface orthogonally, there is in general at each point but one direction on the surface in which

---

* The distinction between the two cases may be further elucidated thus: Select from the doubly infinite system of auxiliary lines those which pierce a given plane in any closed curve, thus forming a tubular surface of which the lines may be called the elements. Then, in the first case, points moving on the tubular surface and cutting the elements orthogonally will describe closed curves; but, in the second case, they will describe spirals.

The forces of a conservative system afford an example of the first or integrable case. For, if X, Y and Z are the components, in the directions of the axes, of a force whose direction and magnitude are functions of x, y and z, the lines of force are those whose differential equations are

$$\frac{dx}{X} = \frac{dy}{Y} = \frac{dz}{Z}.$$

The equation

$$Xdx + Ydy + Zdz = 0$$

will be satisfied by a particle moving perpendicularly to the lines of force, so that no work is done upon it by the force; and this equation is integrable, the integral $V = C$ being the equation of a system of *level* surfaces to which the lines of force are everywhere normal.

the point can move perpendicularly to the auxiliary lines. We thus have a singly infinite system of lines on the given surface, for the solution of the restricted problem.

262. In a general solution the assumed surface, as, for example, the cylindrical surface represented by equation (1), Art. 257, must be capable of passing through the line in space representing any particular solution; and, the surface being thus properly determined, the line in question will be a member of the singly infinite system determined upon the surface by the additional integral equation found.

The peculiarity of the general solution of Art. 258 is that the assumed surface $V = \phi(z)$ is made up of elements which are themselves particular solutions of a certain class. We still have a singly infinite system of particular solutions upon the assumed surface, namely, the elements just mentioned. But upon each surface there is in addition the unique solution determined by equation (6). The points on the line thus determined are exceptions to the general rule, mentioned in the preceding article, that at each point there is but one direction on the surface in which a point can move perpendicularly to the auxiliary lines. The line is, in fact, the locus of the points at which the auxiliary lines pierce the surface orthogonally.

Examples XX.

Solve the following integrable equations : —

1. $2(y+z)dx + (x+3y+2z)dy + (x+y)dz = 0$,
$$(x+y)^2(y+z) = c.$$

2. $(y-z)dx + 2(x+3y-z)dy - 2(x+2y)dz = 0$,
$$(x+2y)(y-z)^2 = c.$$

3. $(a-z)(ydx + xdy) + xydz = 0$, $\qquad xy = c(z-a)$.

4. $(y+a)^2 dx + z dy - (y+a) dz = 0,$ $z = (x+c)(y+a).$

5. $(ay-bz) dx + (cz-ax) dy + (bx-cy) dz = 0,$
$$(ax-cz) = C(ay-bz).$$

6. $dx + dy + (x+y+z+1) dz = 0,$ $(x+y+z)e^z = c.$

7. $(y^2+yz) dx + (xz+z^2) dy + (y^2-xy) dz = 0,$
$$y(x+z) = c(y+z).$$

8. $(x^2+z^2)(x dx + y dy + z dz) + (x^2+y^2+z^2)^{\frac{1}{2}}(z dx - x dz) = 0,$
$$(x^2+y^2+z^2)^{\frac{1}{2}} + \tan^{-1}\frac{x}{z} = c.$$

9. $2(2y^2+yz-z^2) dx + x(4y+z) dy + x(y-2z) dz = 0,$
$$x^2(y+z)(2y-z) = c.$$

10. $(x^2y - y^3 - y^2z) dx + (xy^2 - x^3 - x^2z) dy + (xy^2 + x^2y) dz = 0,$
$$\frac{x+z}{y} + \frac{y+z}{x} = c.$$

11. $(2x^2 + 2xy + 2xz^2 + 1) dx + dy + 2z dz = 0,$
$$e^{x^2}(y+z^2+x) + c = 0.$$

12. $(2x + y^2 + 2xt - z) dx + 2xy dy - x dz + x^2 dt = 0,$
$$x^2 + xy^2 + x^2t - xz = c.$$

13. $t(y+z) dx + t(y+z+1) dy + t dz - (y+z) dt = 0,$
$$(y+z)e^{x+y} = ct.$$

14. $z(y+z) dx + z(u-x) dy + y(x-u) dz + y(y+z) du = 0,$
$$(y+z)(u+c) + z(x-u) = 0.$$

15. Find the equation which expresses the solution of
$$dz = ay dx + b dy$$
when we assume $y = f(x)$.
$$z = a \int f(x) dx + bf(x) + C.$$

16. Find the equation which determines upon the ellipsoid

$$\frac{x^2}{a^2} + \frac{y^2}{b^2} + \frac{z^2}{c^2} = 1$$

the lines which satisfy

$$xdx + ydy + c\left(1 - \frac{x^2}{a^2} - \frac{z^2}{c^2}\right) dz = 0.$$

$$x^2 + y^2 + z^2 = C^2.$$

17. Find the equations which determine upon the sphere

$$x^2 + y^2 + z^2 = k^2$$

the lines which satisfy

$$\{x(x-a) + y(y-b)\} dz = (z-c)(xdx + ydy).$$

$$z = C, \text{ and } ax + by + cz = k^2.$$

18. Show that, for the differential equation of Ex. 17, the auxiliary system of lines consists of vertical circles, and verify geometrically the results.

19. Give the general solution in Monge's form of the equation

$$zdx + xdy + ydz = 0.$$

$$y + z \log x = \phi(z), \quad x\phi'(z) + y = x \log x.$$

20. Find a general solution of

$$ydx = (x-z)(dy - dz).$$

$$y - z = \phi(x), \quad y = (x-z)\phi'(x).$$

CHAPTER XI.

PARTIAL DIFFERENTIAL EQUATIONS OF THE FIRST ORDER.

XXI.

Equations involving a Single Partial Derivative.

263. An equation of the form

$$Pdx + Qdy + Rdz = 0 \quad \ldots \ldots (1)$$

which satisfies the condition of integrability is sometimes called a *total differential equation*, because it gives the total differential of one of the variables regarded as a function of the other two. Thus, if x and y be the independent variables, the equation gives

$$dz = -\frac{P}{R}dx - \frac{Q}{R}dy,$$

or, in the notation of partial derivatives,

$$\frac{dz}{dx} = -\frac{P}{R}, \quad \ldots \ldots (2)$$

and

$$\frac{dz}{dy} = -\frac{Q}{R}; \quad \ldots \ldots (3)$$

that is to say, we have each of the partial derivatives of z given in the form of a function of x, y and z.

An equation of the form (2) or (3), giving the value of a single partial derivative, or more generally an equation giving a relation between the several partial derivatives of a function of two or more independent variables, is called a *partial differential equation*.

264. To solve a partial differential equation of the simple form (2), it is only necessary to treat it as an ordinary differential equation between x and z, y being regarded as constant, and an unknown function of y taking the place of the constant of integration. The process is the same as that of solving the total differential equation, see Art. 246, except that we have no means of determining the function of y, which accordingly remains arbitrary. Thus the general solution of the equation contains an arbitrary function.

Equations of the First Order and Degree.

265. Denoting the partial derivatives of z by p and q, thus

$$p = \frac{dz}{dx}, \qquad q = \frac{dz}{dy},$$

a partial differential equation of the first order, in which z is the dependent and x and y the independent variables, is a relation between p, q, x, y and z. A relation between x, y and z is a *particular integral*, when the values which it and its derived equations determine for z, p and q in terms of x and y satisfy the given equation identically. We shall find that, as in the case of the simple class of equations considered in the preceding article, the most general solution or *general integral* contains an arbitrary function.

266. The equation of the first order and degree may be written in the form

$$Pp + Qq = R, \quad \ldots \ldots \ldots \quad (1)$$

where P, Q and R are functions of x, y and z. This is sometimes called the *linear* equation, the term linear, in this case, referring only to p and q.

Let
$$u = a, \quad \ldots \ldots \ldots (2)$$

in which u is a function of x, y and z, and a is a constant, be an integral of equation (1). Taking derivatives with respect to x and y, we have

$$\frac{du}{dx} + \frac{du}{dz} p = 0, \quad \text{and} \quad \frac{du}{dy} + \frac{du}{dz} q = 0;$$

and substituting the values of p and q, hence derived in equation (1), we obtain

$$P\frac{du}{dx} + Q\frac{du}{dy} + R\frac{du}{dz} = 0. \quad \ldots \ldots (3)$$

Therefore, if $u = a$ is an integral of equation (1), u is a function satisfying equation (3),* and conversely.

But we have seen in Art. 231 that this equation is satisfied by the function u when $u = a$ is an integral of the system of ordinary differential equations,

$$\frac{dx}{P} = \frac{dy}{Q} = \frac{dz}{R}. \quad \ldots \ldots (4)$$

Hence every integral of the system (4) is also an integral of equation (1).

Now, it was shown in Art. 232, that if

$$u = a \quad \text{and} \quad v = b$$

---

* It follows from the definition of an integral that this equation is either an identity, or becomes such when z is eliminated from it by means of equation (2); but, since it does not contain the constant a which occurs in equation (2), the former alternative must be the correct one.

are two independent integrals of the system (4), the equation

$$f(u, v) = C$$

includes all possible integrals of the system. Hence this equation, in which f is an arbitrary function, is the general integral of equation (1). It is unnecessary to retain an arbitrary constant since f is arbitrary; in fact, solving for u, the equation may be written in the form

$$u = \phi(v),$$

which expresses the relation between x, y and z with equal generality.

Thus, to solve the linear equation (1), we find two independent integrals of the system (4) in the forms $u = a$, $v = b$, and then put $u = \phi(v)$, where ϕ is an arbitrary function. This is known as *Lagrange's* solution.

267. It is readily seen that we can derive in like manner the general integral of the linear partial differential equation containing more than two independent variables. Thus, the equation being

$$P_1\frac{dz}{dx_1} + P_2\frac{dz}{dx_2} + \ldots + P_n\frac{dz}{dx_n} = R, \quad \ldots \ldots (1)$$

the auxiliary system is

$$\frac{dx_1}{P_1} = \frac{dx_2}{P_2} = \ldots = \frac{dx_n}{P_n} = \frac{dz}{R}, \quad \ldots \ldots (2)$$

and, if $u_1 = c_1$, $u_2 = c_2$, ..., $u_n = c_n$ are independent integrals of this system, the general integral of equation (1) may be written

$$f(u_1, u_2, \ldots u_n) = 0, \quad \ldots \ldots \ldots (3)$$

where f is an arbitrary function. If an insufficient number of integrals of the system (2) is known, any one of them, or an equation involving an arbitrary function of two or more of the quantities u_1, u_2, ..., u_n constitutes a particular integral of equation (1).

Geometrical Illustration of Lagrange's Solution.

268. The system of ordinary differential equations employed in Lagrange's process are sometimes called *Lagrange's equations*. In the case of two independent variables they represent a doubly infinite system of lines, which may be called the *Lagrangean lines*. We have seen in Art. 235 that every integral of the differential system represents a surface passing through lines of the system, and not intersecting any of them. It follows, therefore, that the partial differential equation

$$Pp + Qq = R$$

is satisfied by the equation of every surface that passes through lines of the system represented by Lagrange's equations

$$\frac{dx}{P} = \frac{dy}{Q} = \frac{dz}{R};$$

and the general integral is the general equation of the surfaces passing through lines of the system.

Given, for example, the equation

$$(mz - ny)p + (nx - lz)q = ly - mx, \quad \ldots \quad (1)$$

for which Lagrange's equations are

$$\frac{dx}{mz - ny} = \frac{dy}{nx - lz} = \frac{dz}{ly - mx}. \quad \ldots \quad (2)$$

The integrals of this system were found, in Art. 230, to be

$$lx + my + nz = a,$$

and

$$x^2 + y^2 + z^2 = b;$$

and, as stated in Art. 236, the lines represented being circles having a fixed line as axis, every integral of the system (2) represents a surface of revolution having the same line as axis. Thus the general integral of equation (1), which is

$$lx + my + nz = \phi(x^2 + y^2 + z^2), \quad \ldots \quad (3)$$

represents all the surfaces of revolution of which the line

$$\frac{x}{l} = \frac{y}{m} = \frac{z}{n}$$

is the axis.

269. It was shown in Art. 260 that, when

$$Pdx + Qdy + Rdz = 0 \quad \ldots \quad (1)$$

is the differential equation of a system of surfaces, the system of lines represented by

$$\frac{dx}{P} = \frac{dy}{Q} = \frac{dz}{R} \quad \ldots \quad (2)$$

cuts these surfaces orthogonally. It follows that the surfaces represented by the general integral of

$$Pp + Qq = R,$$

which pass through the lines of the system (2), cut the surfaces of the system (1) orthogonally. Hence, as first shown by Lagrange,* if the equation of a system of surfaces containing one parameter c be put in the form

$$V = c,$$

the surfaces which cut the system orthogonally are all included in

$$u = \phi(v),$$

---

* Œuvres de Lagrange, vol. iv. p. 628; vol. v. p. 560.

where $u = a$ and $v = b$ are two independent integrals of

$$\frac{dx}{\dfrac{dV}{dx}} = \frac{dy}{\dfrac{dV}{dy}} = \frac{dz}{\dfrac{dV}{dz}}.$$

The Complete and General Primitives.

270. If, in an equation containing x, y and z, z be regarded as a function of x and y, we may, by differentiation with respect to x and y, obtain equations involving p and q respectively; and by the combination of the given and the two derived equations we can derive a variety of partial differential equations satisfied by the given equation. If the given equation contains two arbitrary constants, their elimination leads to a definite differential equation of the first order independent of these constants, and of this equation the given equation is called a *complete primitive*.

Given, for example, the equation

$$z = a(x + y) + b. \quad \ldots \ldots \quad (1)$$

By differentiation we have $p = a$, and $q = a$, hence

$$p = q \, . \quad \ldots \ldots \ldots \quad (2)$$

is the only equation of the first order independent of a and b, which can be derived from equation (1). Hence equation (1) is a complete primitive of equation (2). We do not say *the* complete primitive, because the general solution of $p = q$ is

$$z = f(x + y), \quad \ldots \ldots \ldots \quad (3)$$

and therefore any equation of this form containing two arbitrary constants is a complete primitive of $p = q$. In fact, equation (3) gives $p = f'(x + y)$, $q = f'(x + y)$, whence $p = q$. The equation from which a given partial differential equation can be obtained by the elimination of an arbitrary function is called its *general primitive;* thus equation (3) is the general primitive of $p = q$.

271. The most general equation between x, y and z, containing one arbitrary function, may be written in the form

$$f(u, v) = 0, \ldots \ldots \ldots (1)$$

where u and v are given functions of x, y and z. Regarding z as a function of x and y, the derived equations are

$$\frac{df}{du}\left[\frac{du}{dx} + \frac{du}{dz}p\right] + \frac{df}{dv}\left[\frac{dv}{dx} + \frac{dv}{dz}p\right] = 0,$$

and

$$\frac{df}{du}\left[\frac{du}{dy} + \frac{du}{dz}q\right] + \frac{df}{dv}\left[\frac{dv}{dy} + \frac{dv}{dz}q\right] = 0.$$

The result of eliminating the ratio $\frac{df}{du} : \frac{df}{dv}$ may be written in the form

$$\begin{vmatrix} \frac{du}{dx} + p\frac{du}{dz} & \frac{du}{dy} + q\frac{du}{dz} \\ \frac{dv}{dx} + p\frac{dv}{dz} & \frac{dv}{dy} + q\frac{dv}{dz} \end{vmatrix} = 0.$$

Of the four determinants formed by the partial columns, that containing pq as a factor vanishes, and we have

$$\begin{vmatrix} \frac{du}{dx} & \frac{du}{dy} \\ \frac{dv}{dx} & \frac{dv}{dy} \end{vmatrix} + p\begin{vmatrix} \frac{du}{dz} & \frac{du}{dy} \\ \frac{dv}{dz} & \frac{dv}{dy} \end{vmatrix} + q\begin{vmatrix} \frac{du}{dx} & \frac{du}{dz} \\ \frac{dv}{dx} & \frac{dv}{dz} \end{vmatrix} = 0,$$

an equation of the form

$$Pp + Qq = R,$$

§ XXI.]   THE GENERAL PRIMITIVE.   295

in which

$$P = \begin{vmatrix} \dfrac{du}{dy} & \dfrac{du}{dz} \\ \dfrac{dv}{dy} & \dfrac{dv}{dz} \end{vmatrix}, \quad Q = \begin{vmatrix} \dfrac{du}{dz} & \dfrac{du}{dx} \\ \dfrac{dv}{dz} & \dfrac{dv}{dx} \end{vmatrix}, \quad R = \begin{vmatrix} \dfrac{du}{dx} & \dfrac{du}{dy} \\ \dfrac{dv}{dx} & \dfrac{dv}{dy} \end{vmatrix}.$$

It thus appears that the equation of which the general primitive contains a single arbitrary function is linear with respect to p and q.

272. The values of P, Q and R above are called the *Jacobians* of u and v with respect to y and z, z and x, x and y respectively, and are denoted thus,

$$P = \frac{d(u, v)}{d(y, z)}, \quad Q = \frac{d(u, v)}{d(z, x)}, \quad R = \frac{d(u, v)}{d(x, y)}.$$

The Jacobian vanishes when u and v are not independent functions of the variables expressed in the denominator, thus R vanishes if either u or v is a function of z only. Again, P, Q and R all vanish if u is expressible as a function of v. In this last case equation (1) is, in fact, reducible to $v = c$, which contains no arbitrary function.

When P, Q and R are given, the functions u and v must be such that their Jacobians are proportional to P, Q and R. Now, if we put

$$u = a \quad \text{and} \quad v = b,$$

we shall have

$$\frac{du}{dx} dx + \frac{du}{dy} dy + \frac{du}{dz} dz = 0,$$

$$\frac{dv}{dx} dx + \frac{dv}{dy} dy + \frac{dv}{dz} dz = 0:$$

whence, solving for the ratios $dx : dy : dz$, we have

$$\frac{dx}{\frac{d(u,v)}{d(y,z)}} = \frac{dy}{\frac{d(u,v)}{d(z,x)}} = \frac{dz}{\frac{d(u,v)}{d(x,y)}}.$$

Hence we shall have found proper values of u and v if $u = a$ and $v = b$ are integrals of

$$\frac{dx}{P} = \frac{dy}{Q} = \frac{dz}{R}.$$

We have thus another proof of Lagrange's solution of the linear equation.

273. In like manner, if there be n independent variables $x_1, x_2, \ldots, x_n$, and one dependent variable z, we can eliminate the arbitrary function f from the equation

$$f(u_1, u_2 \ldots u_n) = 0,$$

in which $u_1, u_2, \ldots, u_n$ are n independent given functions of the variables. In the result of elimination the coefficient of the products of any two or more of the partial derivatives will vanish, and we shall have an equation linear in these derivatives, that is an equation of the form

$$P_1 p_1 + P_2 p_2 + \ldots + P_n p_n = R.$$

Moreover, each of the coefficients $P_1, P_2, \ldots, P_n$ and R will be the Jacobians of $u_1, u_2, \ldots, u_n$ with respect to n of the variables, and the simultaneous ordinary equations derived from $u_1 = c_1, u_2 = c_2, \ldots, u_n = c_n$ will be

$$\frac{dx_1}{P_1} = \frac{dx_2}{P_2} = \ldots = \frac{dx_n}{P_n} = \frac{dz}{R},$$

where $P_1, P_2, \ldots, P_n$ and R are the same Jacobians.

Examples XXI.

Solve the following partial differential equations:—

1. $y\dfrac{dz}{dy} - 2x - 2z - y = 0$, $\qquad x + y + z = y^2\phi(x)$.

2. $p\sqrt{(y^2 - x^2)} = y$, $\qquad z = y\sin^{-1}\dfrac{x}{y} + \phi(y)$.

3. $lp + mq = 1$, $\qquad z = \dfrac{x}{l} + \phi(ly - mx)$.

4. $p + q = nz$, $\qquad z = e^{ny}\phi(x - y)$.

5. $xp + yq = nz$, $\qquad z = x^n\phi\left(\dfrac{y}{x}\right)$.

6. $yp + xq = z$, $\qquad z = (x + y)\phi(x^2 - y^2)$.

7. $(y^3x - 2x^4)p + (2y^4 - x^3y)q = 9z(x^3 - y^3)$,
$$z = \dfrac{1}{x^3y^3}\phi\left(\dfrac{x}{y^2} + \dfrac{y}{x^2}\right).$$

8. $xzp + yzq = xy$, $\qquad z^2 = xy + \phi\left(\dfrac{y}{x}\right)$.

9. $x^2p - xyq + y^2 = 0$, $\qquad z = \dfrac{y^2}{3x} + \phi(xy)$.

10. $zp + yq = x$, $\qquad x + z = y\phi(x^2 - z^2)$.

11. $xp + zq + y = 0$, $\qquad \tan^{-1}\dfrac{y}{z} = \log x + \phi(y^2 + z^2)$.

12. $(y + z)p + (z + x)q = x + y$,
$$(z - y)\sqrt{(x + y + z)} = \phi\left(\dfrac{z - y}{x - z}\right).$$

13. $x^2p + y^2q = nxy$, $\qquad z = \dfrac{nxy}{y - x}\log\dfrac{y}{x} + \phi\left(\dfrac{y - x}{xy}\right)$.

14. $x(y - z)p + y(z - x)q = z(x - y)$, $\qquad xyz = \phi(x + y + z)$.

15. $p - q = \dfrac{z}{x + y}$, $\qquad (x + y)\log z = x + \phi(x + y)$.

16. $z - xp - yq = a\sqrt{(x^2 + y^2 + z^2)}$,
$$z + \sqrt{(x^2 + y^2 + z^2)} = x^{1-a}\phi\left(\frac{y}{x}\right).$$

17. $(y + x)p + (y - x)q = z$,
$$z = \sqrt{(x^2 + y^2)}\,\phi\left[\tan^{-1}\frac{y}{x} + \tfrac{1}{2}\log(x^2 + y^2)\right].$$

18. $y^2p + xyq = nxz$, $\qquad\qquad z = y^n\phi(x^2 - y^2)$.

19. $xy^2p - y^3q + axz = 0$, $\qquad \log z = -\dfrac{ax}{3y^2} + \phi(xy)$.

20. $(S - x_1)p_1 + (S - x_2)p_2 + \ldots + (S - x_n)p_n = S - z$,
where $S = x_1 + x_2 + \ldots + x_n + z$,
$$\phi\{S^{\frac{1}{n}}(x_1 - z),\ S^{\frac{1}{n}}(x_2 - z),\ \ldots,\ S^{\frac{1}{n}}(x_n - z)\} = 0.$$

21. $x\dfrac{dz}{dx} + y\dfrac{dz}{dy} + t\dfrac{dz}{dt} = az + \dfrac{xy}{t}$,
$$(a - 1)z + \dfrac{xy}{t} = x^a\phi\left(\dfrac{y}{x}, \dfrac{t}{x}\right).$$

22. Find a common integral of the equations
$$py = qx \quad \text{and} \quad px + qy = z.$$
$$z = c\sqrt{(x^2 + y^2)}.$$

23. Show that $x^3 + y^3 + z^3 - 3xyz = r^3$ is a surface of revolution, and find its axis.

24. If $u = 0$ and $v = 0$ are particular integrals of a linear partial differential equation, show that every other integral $\phi = 0$ satisfies the equation
$$\frac{d(\phi, u, v)}{d(x, y, z)} = 0.$$

25. Determine the surfaces which cut orthogonally the system of similar ellipsoids
$$\frac{x^2}{m^2} + \frac{y^2}{n^2} + z^2 = c^2. \qquad \phi\left(\frac{xm^2}{z}, \frac{yn^2}{z}\right) = 0.$$

26. Determine the surfaces of the second order which cut orthogonally the spheres $\quad x^2 + y^2 + z^2 = 2ax$.
$$x^2 + y^2 + z^2 = 2by + 2cz.$$

XXII.

The Non-Linear Equation of the First Order.

274. We have seen in Art. 270 that a partial differential equation of the first order may be derived from a given primitive by the elimination of two arbitrary constants. Such a primitive constitutes a complete integral of the differential equation; but, when the resulting equation is linear, the general solution contains an arbitrary function which imparts a generality infinitely transcending that produced by the presence of arbitrary constants or parameters. The surfaces represented by a complete integral constitute a doubly infinite system of surfaces of the same kind, while the more general class of surfaces represented by the general integral is said to form a *family of surfaces*. Thus, in the example given in Art. 270, the complete integral (1) represents the doubly infinite system of planes parallel to a fixed line; and the general integral (3) represents the family of cylindrical surfaces whose elements are parallel to the same fixed line.

275. The differential equation derived from a complete primitive may be non-linear. For example, if, in the primitive,

$$(x-h)^2 + (y-k)^2 + z^2 = c^2, \quad \ldots \ldots (1)$$

h and k are regarded as arbitrary parameters, the resulting differential equation is

$$z^2(p^2 + q^2 + 1) = c^2, \quad \ldots \ldots (2)$$

which is not linear with respect to p and q. Equation (1) is therefore a complete integral of equation (2). Geometrically it represents a doubly infinite system of equal spheres having their centres in the plane of xy. It will be shown, however, in the following articles, that the geometrical representation of the general integral of a non-linear equation is a family of surfaces equally general with that representing the general integral of a linear equation. But, since it has been shown in Art. 271 that a primitive containing an arbitrary function gives rise in all cases to a linear equation, it is obvious that the general integral of a non-linear differential equation cannot be expressed by a single equation.*

The System of Characteristics.

276. A partial differential equation of the first order, containing two independent variables, is of the form

$$F(x, y, z, p, q) = 0. \qquad (1)$$

Let
$$z = \phi(x, y), \qquad (2)$$
whence
$$p = \frac{d\phi}{dx}, \qquad q = \frac{d\phi}{dy}, \qquad (3)$$

be an integral; then these values of z, p and q satisfy equation (1) identically. If x, y and z be regarded as the coordinates of a point, equation (2) represents a surface. A set of corresponding values of x, y, z, p and q determine not only a point upon the surface, but the direction of the tangent plane at that point, and are said to determine an *element of the surface*. If we permit x and y to vary simultaneously in any manner, the corresponding consecutive elements of surface determine a *linear*

---

* The surfaces of the same family are generated by the motion of a curve in space, when arbitrary relations exist between its parameters. The simplest case is that in which there are but two parameters; the two equations of the curve can then be put in the form $u = c_1$, $v = c_2$; and, if $f(c_1, c_2) = 0$ is the relation between the parameters, $f(u, v) = 0$ is the general equation of the family. This case, therefore, corresponds to the linear differential equation. See Salmon's "Geometry of Three Dimensions," Dublin, 1874, pp 372 *et seq.*

§ XXII.] THE SYSTEM OF CHARACTERISTICS. 301

element of surface; that is, a line upon the surface together with the direction of the tangent plane at each point of the line.

The linear element thus determined upon the surface (2) will in general depend upon the form of the function ϕ; but it will now be shown that, starting from any initial point upon the surface, there exists one linear element which is independent of the form of ϕ, provided only that equation (1) is satisfied, so that every integral surface which passes through the initial element must contain the entire linear element.

277. Let the partial derivatives of F be denoted as follows:

$$\frac{dF}{dx} = X, \quad \frac{dF}{dy} = Y, \quad \frac{dF}{dz} = Z, \quad \frac{dF}{dp} = P, \quad \frac{dF}{dq} = Q.$$

Since z, p and q are functions of x and y, the derivatives of equation (1) with respect to x and y give

$$X + Zp + P\frac{dp}{dx} + Q\frac{dq}{dx} = 0, \quad \ldots \ldots (4)$$

$$Y + Zq + P\frac{dp}{dy} + Q\frac{dq}{dy} = 0. \quad \ldots \ldots (5)$$

Now let x and y vary simultaneously in such a way that

$$\frac{dx}{dt} = P, \quad \frac{dy}{dt} = Q; \quad \ldots \ldots (6)$$

then, because for every point moving in the surface

$$dz = pdx + qdy,$$

we have also

$$\frac{dz}{dt} = pP + qQ. \quad \ldots \ldots \ldots (7)$$

Equations (6) and (7) give

$$\frac{dx}{P} = \frac{dy}{Q} = \frac{dz}{pP + qQ}.$$

The values of p and q in these equations being given in terms of x and y, by equations (3), they form a differential system for the variables x, y and z. Starting from any initial point (x_0, y_0, z_0), this system determines a line in space; and, supposing the initial point to be taken on the surface (2), this line lies upon that surface.

Now, substituting from equation (6), and remembering that

$$\frac{dq}{dx} = \frac{d^2z}{dxdy} = \frac{dp}{dy},$$

equation (4) becomes

$$X + Zp + \frac{dp}{dx}\frac{dx}{dt} + \frac{dp}{dy}\frac{dy}{dt} = 0,$$

whence

$$\frac{dp}{dt} = -X - Zp. \quad \ldots \ldots \ldots (8)$$

In like manner, equation (5) gives

$$\frac{dq}{dt} = -Y - Zq. \quad \ldots \ldots \ldots (9)$$

Equations (6), (7), (8) and (9) now give

$$\frac{dx}{P} = \frac{dy}{Q} = \frac{dz}{pP + qQ} = -\frac{dp}{X + Zp} = -\frac{dq}{Y + Zq}, \quad (10)$$

a complete differential system for the five variables x, y, z, p and q. Starting from any initial element of surface $(x_0, y_0, z_0, p_0, q_0)$, this system determines a linear element of surface, and supposing the initial element to be taken on the surface (2), the entire linear element lies upon that surface.

Now the system (10) is independent of the form of the function ϕ, and the only restriction upon the initial element is that it must satisfy equation (1); it follows that every integral surface which contains the initial element contains the entire linear element. This linear element, depending only upon the form of equation (1), is called a *characteristic* of the partial differential equation. Through every element which satisfies equation (1) there passes a characteristic.*

278. A complete solution of the system (10) consists of four integrals in the form of relations between x, y, z, p and q. Multiplying the terms of the several fractions by X, Y, Z, $-P$ and $-Q$, respectively, we obtain the exact equation $dF = 0$, of which $F = C$ is the integral. But it is obvious that, in order to confine our attention to the characteristics of the given equation, we must take $C = 0$. Thus the original equation is to be taken as one of the integrals of the characteristic system. The other three integrals introduce three arbitrary constants. Hence the characteristics form a triply infinite system.

For example, in the case of the equation given in Art. 275, which may be written

$$F = p^2 + q^2 - \frac{c^2}{z^2} + 1 = 0, \quad \ldots \ldots (1)$$

$X = 0$, $Y = 0$, $Z = \dfrac{2c^2}{z^3}$, $P = 2p$, $Q = 2q$, and the equations of the characteristic are

---

* In like manner, when there are n independent variables, a set of values of $x_1, x_2, \ldots, x_n, z, p_1, p_2, \ldots, p_n$, which satisfies the differential equation, is called an element of its integral, and the consecutive series of elements determined as above are said to form a characteristic. See Jordan's "Cours d'Analyse," Paris, 1887, vol. iii., pp. 318 *et seq.*

$$\frac{dx}{p} = \frac{dy}{q} = \frac{dz}{p^2+q^2} = -\frac{z^3 dp}{c^2 p} = -\frac{z^3 dq}{c^2 q}. \quad \ldots \quad (2)$$

Of this system, equation (1) is an integral; the relation between dp and dq gives a second integral which may be written in the form

$$q = p \tan a. \quad \ldots \ldots \ldots \quad (3)$$

The values of p and q derived from equations (1) and (3) are

$$p = \cos a \frac{\sqrt{(c^2 - z^2)}}{z}, \quad \ldots \ldots \quad (4)$$

$$q = \sin a \frac{\sqrt{(c^2 - z^2)}}{z}, \quad \ldots \ldots \quad (5)$$

and these equations may be taken as two of the integrals, in place of equations (1) and (3). Substituting these values in the relations between dx and dy, dx and dz respectively, we obtain, for the other two integrals,

$$y = x \tan a + a, \quad \ldots \ldots \ldots \quad (6)$$

and

$$(x \sec a + b)^2 = c^2 - z^2. \quad \ldots \ldots \quad (7)$$

These last equations determine, for given values of a, a and b, the characteristic considered merely as a line, and then equations (4) and (5) determine at each point the direction of the element, that is to say, the direction of a plane tangent to every integral surface which passes through the characteristic.

The General Integral.

279. It follows from Art. 277 that every integral surface contains a singly infinite system of characteristics, so that if we make the initial element of a characteristic describe an arbitrary line upon the surface (the linear element of surface along the line determining at each point the values of p_0 and q_0), the locus of the variable characteristic will be the integral surface. Moreover, if we take an arbitrary line in space for the path of the initial point, it is possible so to determine p_0 and q_0 at each point that the characteristic shall generate an integral surface. For this purpose, we must have in the first place,

$$F(x_0, y_0, z_0, p_0, q_0) = 0. \quad \ldots \quad \ldots \quad (1)$$

Again, since the path of the initial point is to lie in the surface, so that

$$dz_0 = p_0 dx_0 + q_0 dy_0,$$

taking the differential equations of the arbitrary curve to be

$$\frac{dx_0}{L} = \frac{dy_0}{M} = \frac{dz_0}{N}, \quad \ldots \quad \ldots \quad (2)$$

we must have

$$N = p_0 L + q_0 M, \quad \ldots \quad \ldots \quad (3)$$

where L, M and N are functions of x_0, y_0 and z_0. Geometrically, this last equation expresses the condition that the initial element must be so taken that the plane tangent to the surface shall contain the line tangent to the arbitrary curve.

The general integral may now be defined as representing the family of surfaces generated by a variable characteristic having its motion thus directed by an arbitrary curve.*

---

* That the surface thus generated is necessarily an integral will be seen in the following articles to result from the existence of a complete integral. The analytical proof requires that it be shown that, for a point moving in the surface, we have always

$$dz = p dx + q dy,$$

where p and q are given by the equations of the characteristic. If the common value of each member of the equations (2) be denoted by $d\tau$, the variation of τ moves

In the case of the linear equation, when the characteristics become the Lagrangean lines, the values of p_0 and q_0 are still those which satisfy equations (1) and (3); but they need not be considered, because there is but one Lagrangean line through each point.

Derivation of a Complete Integral from the Equations of the Characteristic.

280. The four integrals of the characteristic system contain x, y, z, p, q, and three constants. We may therefore obtain, by elimination if necessary, a relation between x, y, z and two of the constants. Every such equation represents, for any fixed values of the constants, a surface passing through a singly infinite system of characteristics, but not in general a system of the kind considered in Art. 279, so that the equation is not in general an integral of the partial differential equation. It will now be

---

the characteristic, and that of t [dt being, as in Art. 277, the common value of each member of equations (10)] moves a point along the characteristic. The motion of a point along the surface then depends upon the two independent variables t and τ. Then, since

$$dz = \frac{dz}{dt}dt + \frac{dz}{d\tau}d\tau, \qquad dx = \frac{dx}{dt}dt + \frac{dx}{d\tau}d\tau, \qquad dy = \frac{dy}{dt}dt + \frac{dy}{d\tau}d\tau,$$

and the equations of the characteristic give

$$\frac{dz}{dt} = p\frac{dx}{dt} + q\frac{dy}{dt},$$

it remains only to prove that

$$\frac{dz}{d\tau} = p\frac{dx}{d\tau} + q\frac{dy}{d\tau},$$

or that

$$\frac{dz}{d\tau} - p\frac{dx}{d\tau} - q\frac{dy}{d\tau} = U = 0.$$

Letting $t = 0$ correspond to the initial point, the condition $dz_0 = p_0 dx_0 + q_0 dy_0$ shows that the corresponding value of U is zero, that is $U_0 = 0$. Consider now the value of $\frac{dU}{dt}$. This is

$$\frac{dU}{dt} = \frac{d^2z}{dt\,d\tau} - \frac{dp}{dt}\frac{dx}{d\tau} - p\frac{d^2x}{dt\,d\tau} - \frac{dq}{dt}\frac{dy}{d\tau} - q\frac{d^2y}{dt\,d\tau}.$$

shown how we may find such an integral, that is to say, since two arbitrary constants occur, a complete integral of the given equation.

Suppose one integral of the characteristic system, in addition to the original equation $F=0$, to have been found. Let α denote the constant of integration introduced, and consider the values of p and q in terms of x, y, z and α determined by these equations. Now, in a complete solution of the characteristic system, each characteristic is particularized by a special value for each of the three constants of integration. We may distinguish those in which α has the special value α_1, as the α_1-characteristics; these constitute a doubly infinite system of linear elements of surface, which together include all the point elements determined by the above-mentioned values of p and q, when the particular value α_1 is assigned to α.

Now these α_1-characteristics lie upon a system of integral surfaces. To show this, consider a transverse plane of refer-

---

But
$$\frac{d^2z}{dt\,d\tau} = \frac{d}{d\tau}\frac{dz}{dt} = \frac{dp}{d\tau}\frac{dx}{dt} + p\frac{d^2x}{d\tau\,dt} + \frac{dq}{d\tau}\frac{dy}{dt} + q\frac{d^2y}{d\tau\,dt};$$

hence
$$\frac{dU}{dt} = \frac{dp}{d\tau}\frac{dx}{dt} + \frac{dq}{d\tau}\frac{dy}{dt} - \frac{dp}{dt}\frac{dx}{d\tau} - \frac{dq}{dt}\frac{dy}{d\tau}.$$

Substituting from the equations of the characteristic, this becomes
$$\frac{dU}{dt} = P\frac{dp}{d\tau} + Q\frac{dq}{d\tau} + X\frac{dx}{d\tau} + pZ\frac{dx}{d\tau} + Y\frac{dy}{d\tau} + qZ\frac{dy}{d\tau};$$

or, since $Zdz + Xdx + Ydy + Pdp + Qdq = 0$,
$$\frac{dU}{dt} = -Z\frac{dz}{d\tau} + pZ\frac{dx}{d\tau} + qZ\frac{dy}{d\tau} = -ZU.$$

The integration of this gives
$$U = Ce^{-\int_0^t Z\,dt},$$

and, putting $t=0$, we have $C = U_0 = 0$; hence, so long as the exponential remains finite, $U=0$, which was to be proved. See Jordan's "Course d'Analyse," vol. iii., p. 323.

ence. This is pierced at each point by one of the a_1-characteristics, and at the point the element, which we may take as the initial element of the characteristic, determines in the plane of reference a direction. If, starting from any position in the plane of reference, the initial point moves in the direction thus defined, it describes a determinate curve in that plane, and the corresponding characteristic generates an integral surface. Varying the initial position in the plane of reference, we have a singly infinite system of curves in that plane, and a singly infinite system of integral surfaces.

We have thus a system of surfaces at every point of which the values of p and q are the values above mentioned which involve a_1. Hence, if these values be substituted in the equation

$$dz = pdx + qdy$$

(which, it will be noticed, is, by Art. 277, one of the differential equations of the characteristic system), we shall have an equation true at every point of this system of surfaces; in other words, we shall have the differential equation of the system.*

The integral of this equation will contain a second constant of integration β; when both constants are regarded as arbitrary, it represents a doubly infinite system of surfaces containing the entire system of characteristics, and is a complete integral.

281. As an illustration, let us resume the example of Art. 278. Substitution of the values of p and q, equations (4) and (5), in $dz = pdx + qdy$, gives

$$\frac{zdz}{\sqrt{(c^2 - z^2)}} = dx \cos a + dy \sin a.$$

---

* It follows that the equation thus found is always integrable. This would, of course, not be generally true if the values of p and q simply satisfied the equation $F = 0$. The early researches in partial differential equations were directed to the discovery of values of p and q which satisfied $F = 0$ and at the same time rendered $dz = pdx + qdy$ integrable. See Art. 294.

whence, integrating, we have

$$z^2 + (x \cos\alpha + y \sin\alpha + \beta)^2 = c^2,$$

which is therefore a complete integral of the given equation

$$z^2(p^2 + q^2 + 1) = c^2.$$

This complete integral represents a right circular cylinder of radius c, having its axis in the plane of xy; and since equation (6), Art. 278, represents a plane perpendicular to the axis, we see that the characteristics in this example are equal vertical circles, with their centres in the plane of xy, regarded as elements of right cylinders.

It follows that the general integral represents the family of surfaces generated by a circle of radius c, moving with its centre in, and its plane normal to, an arbitrary curve in the plane of xy. The surfaces included in the complete integral just found are those described when the arbitrary path of the centre is taken as a straight line.

Relation of the General to the Complete Integral.

282. Since all the integral surfaces which pass through a given characteristic touch one another along the characteristic, and the surfaces included in a complete integral contain all the characteristics, it follows that every integral surface touches at each of its points the surface corresponding to a particular pair of values of α and β in the equation of the complete integral. The series of surfaces which touch a given integral surface corresponds to a definite relation between β and α, say $\beta = \phi(\alpha)$; thus the given integral is the envelope of the system of surfaces selected from the complete integral by putting $\beta = \phi(\alpha)$ and so obtaining an equation containing a single arbitrary parameter.

The equation of the envelope of a system of surfaces represented by such an equation is found in the same manner as that of a system of curves. See Diff. Calc., Art. 365. That is to say, we eliminate the arbitrary parameter from the given equation by means of its derivative with respect to this parameter.

283. For example, in the complete integral found in Art. 281, if α and β are connected by the relation

$$h \cos\alpha + k \sin\alpha + \beta = 0, \quad \ldots \ldots (1)$$

the equation becomes

$$z^2 + [(x-h)\cos\alpha + (y-k)\sin\alpha]^2 = c^2. \quad \ldots (2)$$

Taking the derivative with respect to α, we obtain

$$[(x-h)\cos\alpha + (y-k)\sin\alpha][(y-k)\cos\alpha - (x-h)\sin\alpha] = 0,$$

whence we must have either

$$(x-h)\cos\alpha + (y-k)\sin\alpha = 0, \quad \ldots \ldots (3)$$

or else

$$(y-k)\cos\alpha - (x-h)\sin\alpha = 0. \quad \ldots \ldots (4)$$

The elimination of α from equation (2) by means of equation (3) gives

$$z^2 = c^2, \quad \ldots \ldots \ldots (5)$$

and, in like manner, from equations (2) and (4) we obtain

$$z^2 + (x-h)^2 + (y-k)^2 = c^2. \quad \ldots \ldots (6)$$

Equation (1) expresses the condition that the axis of the cylinder represented by the complete integral shall pass through the fixed point $(h, k, 0)$; accordingly the envelope of the system (2) consists of the planes $z = \pm c$, and the sphere (6) whose centre is

(h, k, o). Regarding h and k as arbitrary, equation (6) is the complete integral from which as a primitive the differential equation was derived in Art. 275.

284. To express the general integral, the relation between the constants in the complete integral must be arbitrary. Thus, the complete integral being in the form

$$f(x, y, z, a, b) = 0, \quad \ldots \ldots (1)$$

we may put $b = \phi(a)$, where ϕ denotes an arbitrary function, and then the general integral is the result of eliminating a between the equations,

$$f[x, y, z, a, \phi(a)] = 0, \quad \ldots \ldots (2)$$

and

$$\frac{d}{da} f[x, y, z, a, \phi(a)] = 0. \quad \ldots \ldots (3)$$

The elimination cannot be performed until the form of ϕ is specified; for, as remarked in Art. 275, the general integral cannot be expressed by a single equation unless the given partial differential equation is linear.

Since the general integral can thus be expressed by the aid of any complete integral, we shall hereafter regard a non-linear partial differential equation as solved when a complete integral is found.

Singular Solutions.

285. There may exist a surface which at each of its points touches one of the surfaces included in the complete integral without passing through the corresponding characteristic. Every element of such a surface obviously satisfies the differential equation, and its equation, not being included in the general integral, is a *singular solution* analogous to those which occur in the case of ordinary differential equations.

An integral surface generated, as in Art. 279, by a moving characteristic will in general touch the surface representing the singular solution along a line. If the surfaces of the complete integral have this character, the singular solution will be a part of the envelope found by the process given in the preceding article, no matter what the form of ϕ may be. In this case, equations (2) and (3), Art. 284, which together determine the ultimate intersection of consecutive surfaces of the system (2), represent a characteristic and also the line of tangency with the singular solution. The former, as a varies, generates a surface belonging to the general integral, and the latter generates the singular solution. Thus, in the example of Art. 283, equation (3) determines upon the cylinder (2) its lines of contact with the planes $z = \pm c$, and equation (4) determines a characteristic.

286. There is, however, when a singular solution exists, a special class of integrals which touch the singular solution in single points, each of these being in fact the envelope of those members of the complete integral which pass through a given point on the singular solution. This class of integrals obviously constitutes a doubly infinite system, and thus forms a complete integral of a special kind. The complete integral (6), Art. 283, is an example.

When
$$f(x, y, z, a, b) = 0$$

is the complete integral of this special kind, the characteristics represented by equations (2) and (3), Art. 284, will, for given values of a and b, all pass through a common point, independently of the form of ϕ, and this point will be upon the singular solution. In particular, the characteristic defined by $f = 0$ and $\frac{df}{da} = 0$ will intersect that defined by $f = 0$ and $\frac{df}{db} = 0$, in a point on the singular solution. Hence, in this case, the singular solution will be the result of eliminating a and b from the three equations,

$$f = 0, \quad \frac{df}{da} = 0, \quad \frac{df}{db} = 0.$$

It is to be noticed, however, that the eliminant of these equations may, as in the case of ordinary differential equations, include certain loci which are not solutions of the differential equation.

287. Since the characteristics which lie upon a surface of the kind considered above, all pass through the point of contact with the singular solution, it follows that the singular solution is the locus of a point such that all the characteristics which pass through it have a common element. At such a point, therefore, the initial element fails to determine the direction of the characteristic. Now, in the equations (10), Art. 277, the ratio $dx:dy$ is indeterminate only when $P=0$ and $Q=0$, or when $P=\infty$ and $Q=\infty$; hence one of these conditions must hold at every point of a singular solution. The former is the more usual case, so that a singular solution generally results from the elimination of p and q from

$$F(x, y, z, p, q) = 0$$

by means of the equations

$$\frac{dF}{dp} = 0 \quad \text{and} \quad \frac{dF}{dq} = 0.$$

It is necessary, however, to ascertain whether the locus thus found is a solution of the differential equation, for the conditions $P=0$, $Q=0$, and $P=\infty$, $Q=\infty$ are satisfied at certain other points besides those situated upon a singular solution; for example, those at which all the characteristics which pass through them touch one another. In the example of Art. 278, $P=0$, $Q=0$ gives the singular solution $z = \pm c$, and $P=\infty$, $Q=\infty$ gives $z^2 = 0$, which is the locus of the last-mentioned points, and not a solution.

Equations Involving p and q only.

288. We proceed to consider certain cases in which a complete integral is readily obtained. In the first place, let the equation be of the form

$$F(p, q) = 0. \qquad (1)$$

In this case, since $X = 0$, $Y = 0$, $Z = 0$, two of the equations [(10), Art. 277] of the characteristic become $dp = 0$ and $dq = 0$; whence

$$p = a \quad \text{and} \quad q = b. \qquad (2)$$

The constants a and b are not independent, for, substituting in equation (1), we have

$$F(a, b) = 0. \qquad (3)$$

Substituting in $dz = p\,dx + q\,dy$, we obtain

$$dz = a\,dx + b\,dy;$$

whence, integrating, we have the complete integral

$$z = ax + by + c, \qquad (4)$$

where a and b are connected by equation (3), and c is a second arbitrary constant.

289. The characteristics in this case are straight lines, and the complete integral (4) represents a system of planes. The general integral is a developable surface. There is no singular solution.

A special class of integrals which may be noticed are the envelopes of those planes belonging to the system (4) which pass through a fixed point.* These are obviously cones, whose

---

* The characteristics which pass through a common point in all cases determine an integral surface. The integrals of this special kind constitute a triply infinite system: we may limit the common point or vertex to a fixed surface (as, for example, in Art. 286, to the singular solution), and still have a complete integral.

elements are the characteristics which pass through the fixed point. For example, if the equation is

$$p^2 + q^2 = m^2,$$

these cones are right circular cones with vertical axes, and their equations are

$$(z - \gamma)^2 = m^2(x - \alpha)^2 + m^2(y - \beta)^2.$$

Equation Analogous to Clairaut's.

290. There is another case in which the characteristics are straight lines; namely, when the equation is of the form

$$z = px + qy + f(p, q) \quad \ldots \quad \ldots \quad (1)$$

In this case, $X = p$, $Y = q$, $Z = -1$, and we have again, for two of the equations of the characteristic, $dp = 0$ and $dq = 0$; whence

$$p = a, \qquad q = b. \quad \ldots \quad \ldots \quad (2)$$

Substituting in $dz = p\,dx + q\,dy$, and integrating, we have the complete integral

$$z = ax + by + c, \quad \ldots \quad \ldots \quad (3)$$

in which the constant c is not independent of a and b; for, substituting the values of p and q, equation (1) becomes

$$z = ax + by + f(a, b), \quad \ldots \quad \ldots \quad (4)$$

which, since it is also one of the integrals of the characteristic system, must be identical with equation (3).

291. The complete integral in this case also represents a system of planes, and the general integral is a developable surface. A singular solution also exists.

For example, let the equation be

$$z = px + qy + k\sqrt{(1 + p^2 + q^2)}; \quad \ldots \ldots (1)$$

the complete integral is

$$z = ax + by + k\sqrt{(1 + a^2 + b^2)}. \quad \ldots \ldots (2)$$

For the singular solution, taking the derivatives with respect to a and b, we have

$$x + \frac{ak}{\sqrt{(1 + a^2 + b^2)}} = 0,$$

and

$$y + \frac{bk}{\sqrt{(1 + a^2 + b^2)}} = 0.$$

These equations give

$$a = \frac{-x}{\sqrt{(k^2 - x^2 - y^2)}}, \qquad b = \frac{-y}{\sqrt{(k^2 - x^2 - y^2)}},$$

and, substituting in equation (2), we have

$$x^2 + y^2 + z^2 = k^2. \quad \ldots \ldots (3)$$

Thus the singular solution represents a sphere, the complete integral (2) its tangent planes, and the general integral the developable surface which touches the sphere along any arbitrary curve.

Equations not Containing x or y.

292. When the independent variables do not explicitly occur, the equation is of the form

$$F(z, p, q) = 0. \quad \ldots \ldots (1)$$

§ XXII.]   EQUATIONS OF SPECIAL FORMS.   317

Here $X = 0$ and $Y = 0$, and the final equation of the characteristic system reduces to
$$\frac{dp}{p} = \frac{dq}{q},$$
whence
$$q = ap. \quad \ldots \ldots \ldots \ldots (2)$$

Substituting in equation (1), we have $F(z, p, ap) = 0$, the solution of which gives for p a value of the form
$$p = \phi(z).$$

Thus, $dz = pdx + qdy$ becomes
$$dz = \phi(z)(dx + ady);$$
whence, integrating, we have the complete integral,
$$x + ay = \int \frac{dz}{\phi(z)} + b. \quad \ldots \ldots (3)$$

The illustrative example of Arts. 278 and 281 is an instance of this form. It will be noticed that the mode of solution leads to a complete integral representing cylindrical surfaces whose elements are parallel to the plane of xy. The equation
$$F(z, 0, 0) = 0,$$
representing certain planes parallel to the plane of xy, will obviously be the singular solution.

Equations of the Form $f_1(x, p) = f_2(y, q)$.

293. When the equation does not explicitly contain z, it may be possible to separate the variables x and p from y and q, thus putting the equation in the form
$$f_1(x, p) = f_2(y, q). \quad \ldots \ldots \ldots (1)$$

In this case, we have $Z = 0$, $X = \dfrac{df_1}{dx}$, $P = \dfrac{df_1}{dp}$, and the equations of the characteristic give for the relation between dx and dp,

$$\frac{df_1}{dx} dx + \frac{df_1}{dp} dp = 0.$$

Integrating, we have $f_1(x, p) = a$, and from equation (1),

$$f_1(x, p) = f_2(y, q) = a. \quad \dots \quad \dots \quad (2)$$

Solving these equations for p and q, we have values of the form

$$p = \phi_1(x, a), \qquad q = \phi_2(y, a),$$

and $dz = pdx + qdy$ becomes

$$dz = \phi_1(x, a) dx + \phi_2(y, a) dy,$$

whence we derive the complete integral,

$$z = \int \phi_1(x, a) dx + \int \phi_2(y, a) dy + b.$$

For example, let the given equation be

$$xp^2 + yq^2 = 1.$$

Putting

$$xp^2 = 1 - yq^2 = a,$$

we have

$$p = \frac{\sqrt{a}}{\sqrt{x}}, \qquad q = \frac{\sqrt{(1-a)}}{\sqrt{y}},$$

and, integrating $dz = pdx + qdy$, we obtain the complete integral.

$$z = 2\sqrt{a}\sqrt{x} + 2\sqrt{(1-a)}\sqrt{y} + b.$$

Change of Form in the Equations of the Characteristic.

294. If we make any algebraic change in the form of the equation

$$F(x, y, z, p, q) = 0,$$

the equations of the characteristic (10), Art. 277, will be altered. The changes, however, will be merely such modifications as might be produced by means of the equation $F=0$ itself.* In particular, the form assumed when the equation is first solved for q may be noticed. Suppose the equation to be

$$q = \phi(x, y, z, p). \quad \ldots \ldots \ldots (1)$$

Then $F = q - \phi(x, y, z, p)$, whence $X = -\dfrac{d\phi}{dx},\ Y = -\dfrac{d\phi}{dy},\ Z = -\dfrac{d\phi}{dz},$ $P = -\dfrac{d\phi}{dp},$ and $Q = 1$. Putting q in the place of ϕ in the partial derivatives, and omitting the member containing dq, the equations of the characteristic become

$$\dfrac{dx}{-\dfrac{dq}{dp}} = dy = \dfrac{dz}{q - p\dfrac{dq}{dp}} = \dfrac{dp}{\dfrac{dq}{dx} + p\dfrac{dq}{dz}}, \quad \ldots (2)$$

a complete system for the four variables x, y, z and p, q being the function of these variables, given by equation (1). These equations may be deduced from the consideration that the values of p and q derived from one of their integrals combined with equation (1) should render $dz = pdx + qdy$ integrable.†

---

* The complete solution of the characteristic system involving four arbitrary constants (see Art. 278) would indeed be changed, but not the special solution in which $F = 0$ is taken as one of the integrals.

† See Boole's "Differential Equations," London, 1865, p. 336.

295. As an illustration, let us take the equation

$$z = pq, \quad \text{or} \quad q = \frac{z}{p}. \quad \ldots \ldots (1)$$

Equations (2) of the preceding article become

$$\frac{p^2 dx}{z} = dy = \frac{p\,dz}{2z} = dp. \quad \ldots \ldots (2)$$

Of these the most obvious integral is

$$p = y + a;$$

whence $dz = p\,dx + q\,dy$ becomes

$$dz = (y+a)dx + \frac{z\,dy}{y+a},$$

from which we derive the complete integral

$$z = (y+a)(x+b). \quad \ldots \ldots (3)$$

The equations of the characteristic derived from the more symmetrical form of the equation

$$F = pq - z = 0$$

are

$$\frac{dx}{q} = \frac{dy}{p} = \frac{dz}{2pq} = \frac{dp}{p} = \frac{dq}{q}, \quad \ldots \ldots (4)$$

which are readily seen to be equivalent to equations (2). If the final equation of the system (4) be used, as in the process of Art. 292, to determine p and q, we shall have

$$p = \frac{\sqrt{z}}{a}, \quad q = a\sqrt{z},$$

giving

$$4z = \left(\frac{x}{a} + ay + \beta\right)^2, \quad \ldots \ldots (5)$$

another complete integral of the equation $z = pq$.

Transformation of the Variables.

296. A partial differential equation may sometimes be reduced by transformation of the variables to one of the forms for which complete integrals have been given in Arts. 288, 290, 292 and 293. The simplest transformation is that in which each variable is replaced by an assumed function of itself. The choice of the new variable will be suggested by the form of the given equation.

Let
$$\xi = \phi(x), \qquad \eta = \psi(y), \qquad \zeta = f(z),$$
then
$$d\zeta = f'(z)dz = f'(z)\left(\frac{dz}{d\xi}d\xi + \frac{dz}{d\eta}d\eta\right)$$
$$= \frac{f'(z)dz}{\phi'(x)dx}d\xi + \frac{f'(z)dz}{\psi'(y)dy}d\eta.$$

Hence, denoting the partial derivatives of ζ with respect to ξ and η by p' and q', their expressions in terms of x, y and z are the same as if they were ordinary derivatives.

For example, the equation
$$x^2 p^2 + y^2 q^2 = z^2 \quad \dots \dots \dots (1)$$
may be written
$$\left(\frac{xdz}{zdx}\right)^2 + \left(\frac{ydz}{zdy}\right)^2 = 1.$$

Putting $\dfrac{dx}{x} = d\xi$, $\dfrac{dy}{y} = d\eta$, $\dfrac{dz}{z} = d\zeta$, whence $\xi = \log x$, $\eta = \log y$ and $\zeta = \log z$, the equation becomes
$$p'^2 + q'^2 = 1. \quad \dots \dots \dots (2)$$

The complete integral of this equation is, by Art. 288,
$$\zeta = a\xi + b\eta + c,$$

where $a^2 + b^2 = 1$; hence, putting $a = \cos a$, $b = \sin a$, the complete integral of equation (1) is

$$\log z = \cos a \log x + \sin a \log y + c,$$

or

$$z = Cx^{\cos a} y^{\sin a}.$$

297. In the following example the new independent variables are functions of both of the old ones. Given

$$(x^2 + y^2)(p^2 + q^2) = 1. \quad \dots \quad \dots \quad (1)$$

Using the formulae connecting rectangular with polar coordinates,

$$x = r \cos \theta, \qquad y = r \sin \theta,$$

whence

$$r^2 = x^2 + y^2, \qquad \theta = \tan^{-1} \frac{y}{x},$$

we have

$$p = \frac{dz}{dx} = \frac{dz}{dr} \cos \theta - \frac{dz}{d\theta} \frac{\sin \theta}{r},$$

$$q = \frac{dz}{dy} = \frac{dz}{dr} \sin \theta + \frac{dz}{d\theta} \frac{\cos \theta}{r}.$$

Substituting, equation (1) becomes

$$r^2 \left(\frac{dz}{dr}\right)^2 + \left(\frac{dz}{d\theta}\right)^2 = 1;$$

or, putting $d\rho = \dfrac{dr}{r}$,

$$\left(\frac{dz}{d\rho}\right)^2 + \left(\frac{dz}{d\theta}\right)^2 = 1.$$

Hence the integral is

$$z = \rho \cos a + \theta \sin a + \beta$$
$$= \tfrac{1}{2} \cos a \log (x^2 + y^2) + \sin a \tan^{-1} \frac{y}{x} + \beta.$$

The same complete integral may be found directly by the method of characteristics (see Ex. 20).

Examples XXII.

Find complete integrals for the following partial differential equations:—

1. $pq = 1$, $\qquad z = ax + \dfrac{y}{a} + b.$

2. $\sqrt{p} + \sqrt{q} = 2x$, $\qquad z = \tfrac{1}{6}(2x-a)^3 + a^2 y + b.$

3. $p^2 - q^2 = 1$, $\qquad z = x \sec a + y \tan a + b.$

4. $z = px + qy + pq$, $\qquad z = ax + by + ab;$
singular solution, $z = -xy$.

5. $q = xp + p^2$, $\qquad z = axe^y + \tfrac{1}{2}a^2 e^{2y} + b.$

6. $y^2 p^2 - x^2 q^3 = x^2 y^2$, $\qquad z = \tfrac{1}{2}ax^2 + \tfrac{3}{8}(a^2-1)^{\frac{1}{3}}y^{\frac{8}{3}} + b.$

7. $p^2 + q^2 = x + y$, $\qquad z = \tfrac{2}{3}(x+a)^{\frac{3}{2}} + \tfrac{2}{3}(y-a)^{\frac{3}{2}} + b.$

8. $q = 2yp^2$, $\qquad z = ax + a^2 y^2 + b.$

9. $z = px + qy - np^{\frac{1}{n}}q^{\frac{1}{n}}$, $\qquad z = ax + by - na^{\frac{1}{n}}b^{\frac{1}{n}};$
singular solution, $z = (2-n)(xy)^{\frac{1}{2-n}}$.

10. $p^2 - q^2 = \dfrac{x-y}{z}$, $\qquad z^{\frac{3}{2}} = (x+a)^{\frac{3}{2}} + (y+a)^{\frac{3}{2}} + b.$

11. $p = (qy+z)^2$, $\qquad yz = ax + 2\sqrt{(ay)} + b.$

12. $p^2 + q^2 - 2px - 2qy + 1 = 0$,
$$2z = x^2 + y^2 + x\sqrt{(x^2+a)} + y\sqrt{(y^2-1-a)}$$
$$+ \log \dfrac{[x+\sqrt{(x^2+a)}]^a}{[y+\sqrt{(y^2-1-a)}]^{1+a}} + b.$$

13. Denoting $x + ay$ by t, find a complete integral of Ex. 12 in the form
$$t(1+a^2)z = t^2 + t\sqrt{(t^2-1-a^2)} - (1+a^2)\log[t+\sqrt{(t^2-1-a^2)}] + \beta.$$

14. $(p+q)(px+qy) = 1$, $\qquad \sqrt{(1+a)}z = 2\sqrt{(x+ay)} + b.$

15. $x^2y^3z^3pq^2 = 1$, $\qquad \frac{1}{2}z^2 = -\frac{1}{a^2x} - \frac{2a}{\sqrt{y}} + b.$

16. $x^2y^3z^{-3}p^2q = 1$, $\qquad \log\frac{x^a}{bz} = \frac{1}{2a^2y^2}.$

17. $p^2 - y^2q = y^2 - x^2$,

$$z = \frac{a^2}{2}\sin^{-1}\frac{x}{a} + \frac{x\sqrt{(a^2-x^2)}}{2} - \frac{a^2}{y} - y + b.$$

18. Find three complete integrals of

$$pq = px + qy.$$

$1°\quad 2z = \left(\frac{x}{a} + ay\right)^2 + \beta.$

$2°\quad z = xy + y\sqrt{(x^2 - a^2)} + b.$

$3°\quad z = xy + x\sqrt{(y^2 + a'^2)} + b'.$

19. Show directly, by comparison of the values of z, p and q, that a surface included in the integral $2°$ can be found touching at any given point a given surface included in the integral $1°$; and that the relation

$$\beta = 2b - \frac{a^2}{a^2}$$

will then exist between the constants. Hence derive one integral from the other, as in Art. 283. Also show that the similar relations for the other pairs of integrals are

$\qquad \beta = 2b' + a'^2a^2,\qquad$ and $\qquad b - b' = aa'.$

20. Show that $xq - yp = a$ is an integral of the characteristic system for the equation

$$(x^2 + y^2)(p^2 + q^2) = 1;$$

and thence derive the complete integral given in Art. 297.

21. Solve, by means of the transformations $xy = \xi$, $x + y = \eta$, the equation

$$(y - x)(qy - px) = (p - q)^2.$$

$\qquad z = axy + c^2(x+y) + b.$

22. $(x^2 - y^2)pq - xy(p^2 - q^2) = 1.$

$$z = \tfrac{1}{2}a\log(x^2 + y^2) + \frac{1}{c}\tan^{-1}\frac{y}{x} + b.$$

23. Show that the equations of the characteristic passing through (α, β, γ) in the case of the equation

$$p^2 + q^2 = m^2,$$

Art. 289, are

$$\frac{x-\alpha}{a} = \frac{y-\beta}{b} = \frac{z-\gamma}{m^2},$$

where $a^2 + b^2 = m^2$; and thence derive the special integral given in that article.

24. Deduce, in like manner, the integral formed by characteristics passing through (h, k, l) for the equation

$$p^2 + q^2 = \frac{c^2}{z^2} - 1.$$

$$(x-h)^2 + (y-k)^2 = [\sqrt{(c^2-l^2)} - \sqrt{(c^2-z^2)}]^2.$$

25. Show that when the complete integral is of the form

$$au + bv + w = 0, \quad \ldots \ldots \ldots (1)$$

where u, v and w are rational functions of x, y and z, the elimination can be performed, giving the general integral

$$\phi\left(\frac{u}{w}, \frac{v}{w}\right) = 0, \quad \ldots \ldots \ldots (2)$$

a homogeneous equation in u, v, w. Accordingly, show that the equation arising from equation (1) as a primitive is the linear equation $Pp + Qq = R$, where

$$P = u\frac{d(v, w)}{d(y, z)} + v\frac{d(w, u)}{d(y, z)} + w\frac{d(u, v)}{d(y, z)},$$

with similar expressions for Q and R, and that putting $u_1 = \dfrac{u}{w}$, $v_1 = \dfrac{v}{w}$, these values of P, Q and R agree with those derived from the general primitive in Art. 271.

CHAPTER XII.

PARTIAL DIFFERENTIAL EQUATIONS OF HIGHER ORDER.

XXIII.

Equations of the Second Order.

298. WE have seen that the general solution of a partial differential equation of the first order, containing two independent variables, involves an arbitrary function, although it is not possible to express the solution by a single equation except when the differential equation is linear with respect to p and q. We might thus be led to expect that the general solution of an equation of the second order could be made to depend upon two arbitrary functions. But this is not generally the case. No complete theory of the nature of a solution has yet been developed, although in certain cases the general solution is expressible by an equation containing two arbitrary functions. We shall consider these cases in the present section, and in the next, the important class of linear equations with constant coefficients, for which in some cases a solution of the equation of the nth order containing n arbitrary functions can be obtained.

The Primitive containing Two Arbitrary Functions.

299. If we consider on the other hand the question of the differential equation arising from a given primitive by the elimination of two arbitrary functions, we shall find that it is only in

§ XXIII.] TWO ARBITRARY FUNCTIONS. 327

certain cases that the elimination can be performed without introducing derivatives of an order higher than the second.

The general equation containing two arbitrary functions may be written in the form

$$f[x, y, z, \phi(u), \psi(v)] = 0,$$

in which u and v are given functions of x, y and z. The two derived equations

$$\frac{df}{dx} = 0, \qquad \frac{df}{dy} = 0,$$

will contain $\phi'(u)$ and $\psi'(v)$, two new unknown quantities to be eliminated. There will be three derived equations of the second order

$$\frac{d^2f}{dx^2} = 0, \qquad \frac{d^2f}{dx\,dy} = 0, \qquad \frac{d^2f}{dy^2} = 0,$$

containing two new unknown quantities, $\phi''(u)$ and $\psi''(v)$. We have thus in all six equations containing six unknown quantities. The elimination, therefore, cannot in general be effected.*

300. Suppose, however, that the original equation can be put in the form

$$w = \phi(u) + \psi(v); \quad \cdots \cdots \quad (1)$$

then the two derived equations of the first order,

$$\frac{dw}{dx} + \frac{dw}{dz} p = \phi'(u)\left(\frac{du}{dx} + \frac{du}{dz} p\right) + \psi'(v)\left(\frac{dv}{dx} + \frac{dv}{dz} p\right), \quad (2)$$

$$\frac{dw}{dy} + \frac{dw}{dz} q = \phi'(u)\left(\frac{du}{dy} + \frac{du}{dz} q\right) + \psi'(v)\left(\frac{dv}{dy} + \frac{dv}{dz} q\right), \quad (3)$$

are independent of ϕ and ψ. These, with the three derived

---

* If we proceed to the third derivatives, we shall have ten equations and eight quantities to be eliminated, so that two equations of the third order could be found which would be satisfied by the given primitive.

equations of the second order, will constitute five equations containing the four quantities ϕ', ψ', ϕ'', ψ''. These quantities may therefore be eliminated, the result being an equation of the second order.

There is another way in which the elimination may be effected. Let one of the unknown quantities, say ψ', be eliminated between equations (2) and (3); we shall then have a single equation containing ϕ'. From this equation and its two derived equations we can eliminate ϕ' and ϕ''. It is to be noticed that in this last process we meet with an *intermediate equation* of the first order, containing one arbitrary function.

301. Another case in which the elimination can be performed occurs when the primitive is of the form

$$w = \phi(u) + v\psi(u), \quad \ldots \ldots (1)$$

in which we have two arbitrary functions of the same given function of x, y and z. In this case the derived equations take the form

$$\left(\frac{dw}{dx}\right) = \phi'(u)\left(\frac{du}{dx}\right) + v\psi'(u)\left(\frac{du}{dx}\right) + \psi(u)\left(\frac{dv}{dx}\right), \ldots (2)$$

$$\left(\frac{dw}{dy}\right) = \phi'(u)\left(\frac{du}{dy}\right) + v\psi'(u)\left(\frac{du}{dy}\right) + \psi(u)\left(\frac{dv}{dy}\right), \ldots (3)$$

in which $\left(\frac{dw}{dx}\right)$, etc., are written in place of $\frac{dw}{dx} + \frac{dw}{dz}\dot{p}$, etc.

Multiplying equations (2) and (3) by $\left(\frac{du}{dy}\right)$ and $\left(\frac{du}{dx}\right)$ respectively, and subtracting the results, $\phi'(u)$ and $\psi'(u)$ are eliminated together, and we have again an intermediate equation of the first order containing one arbitrary function.*

---

* The cases considered in this and the preceding article are not the only ones in which an intermediate equation of the first order can arise. See, for instance, the example given in Art. 311.

The Intermediate Equation of the First Order.

302. The preceding articles indicate two cases in which an intermediate equation of the first order may arise from a primitive. We have now to consider, on the other hand, the form of the differential equations arising from an intermediate equation of the form

$$u = \phi(v), \quad \ldots \ldots \ldots (1)$$

where u and v now denote given functions of x, y, z, p and q,[*] and ϕ is an arbitrary function. Denoting the second derivatives of z by r, s and t, thus

$$r = \frac{d^2z}{dx^2}, \qquad s = \frac{d^2z}{dxdy}, \qquad t = \frac{d^2z}{dy^2},$$

the two derived equations are

$$\frac{du}{dx} + \frac{du}{dz}p + \frac{du}{dp}r + \frac{du}{dq}s = \phi'(v)\left(\frac{dv}{dx} + \frac{dv}{dz}p + \frac{dv}{dp}r + \frac{dv}{dq}s\right),$$

$$\frac{du}{dy} + \frac{du}{dz}q + \frac{du}{dp}s + \frac{du}{dq}t = \phi'(v)\left(\frac{dv}{dy} + \frac{dv}{dz}q + \frac{dv}{dp}s + \frac{dv}{dq}t\right);$$

and the result of eliminating $\phi'(v)$ may be written in the form

$$\begin{vmatrix} \dfrac{du}{dx} + p\dfrac{du}{dz} + r\dfrac{du}{dp} + s\dfrac{du}{dq} & \dfrac{du}{dy} + q\dfrac{du}{dz} + s\dfrac{du}{dp} + t\dfrac{du}{dq} \\ \dfrac{dv}{dx} + p\dfrac{dv}{dz} + r\dfrac{dv}{dp} + s\dfrac{dv}{dq} & \dfrac{dv}{dy} + q\dfrac{dv}{dz} + s\dfrac{dv}{dp} + t\dfrac{dv}{dq} \end{vmatrix} = 0. \quad (2)$$

Of the sixteen determinants formed by the partial columns,

---

[*] In the cases considered in Arts. 300 and 301, the function v in the intermediate equation does not contain p and q, but we here include all cases in which an intermediate equation of the first order can exist.

those containing rs and st vanish, and the remaining terms of the second degree in r, s, t may be written

$$(rt - s^2) \begin{vmatrix} \dfrac{du}{dp} & \dfrac{du}{dq} \\ \dfrac{dv}{dp} & \dfrac{dv}{dq} \end{vmatrix} \quad \text{or} \quad (rt - s^2) \dfrac{d(u, v)}{d(p, q)}.$$

The equation will therefore be of the form

$$Rr + Ss + Tt + U(rt - s^2) = V. \quad \ldots \quad (3)$$

303. When a given equation of the second order admits of a primitive containing two arbitrary functions, the intermediate equation of the first order is an *intermediate integral* analogous to the first integrals of ordinary differential equations of the second order. It follows from the preceding article that an intermediate integral will exist only when the given equation is of the form (3), and when u and v can be so determined as to make the functions R, S, T, U and V defined by the development of equation (2) proportional to the given coefficients. This imposes four conditions upon the two quantities u and v; hence two identical relations must exist between the coefficients, in order that an intermediate integral may be possible.

Successive Integration.

304. When an intermediate integral can be found, the final integral is derived from it by the methods given in the preceding chapter for equations of the first order; the second integration introducing a second arbitrary function.

In some simple cases it is obvious that an intermediate equation can be obtained by direct integration. Thus, when the equation contains derivatives with respect to one only of the independent variables, we may treat it as an ordinary differential equation of the second order; taking care only to introduce arbitrary functions of the other variable in the place of constants of integration. Given, for example, the equation

$$xr - p = xy,$$

which may be written

$$\frac{xdp - pdx}{x^2} = \frac{ydx}{x}.$$

Integrating,

$$\frac{p}{x} = y \log x + \phi(y),$$

or

$$p = yx \log x + x\phi(y);$$

and, integrating again,

$$z = y(\tfrac{1}{2}x^2 \log x - \tfrac{1}{4}x^2) + \tfrac{1}{2}x^2\phi(y) + \psi(y),$$

or, putting $\phi(y)$ in place of the function $\tfrac{1}{2}\phi(y) - \tfrac{1}{4}y$,

$$z = \tfrac{1}{2}yx^2 \log x + x^2\phi(y) + \psi(y).$$

305. Again, an equation which does not contain t may be exact * with reference to x, y being regarded as constant. Given, for example, the equation

$$p + r + s = 1;$$

integrating, we have

$$z + p + q = x + \phi(y).$$

---

* The equation might also be such as to become exact with respect to the four variables p, q, z and x, by means of a factor. For this purpose three conditions of integrability would have to be satisfied; see Art. 252. This is the number of conditions we should expect, since by Art. 303 two must be fulfilled to render an intermediate integral possible, and one more is necessary to express that in that integral $v = y$.

For this linear equation of the first order, Lagrange's equations are

$$dx = dy = \frac{dz}{x - z + \phi(y)},$$

of which the first gives

$$x - y = a,$$

and this converts the second into

$$\frac{dz}{dy} + z = a + y + \phi(y),$$

of which the integral is

$$e^y z = ae^y + \int [y + \phi(y)]e^y dy + b.$$

Hence, making $b = \psi(a)$, we have for the final integral

$$e^y z = e^y x - e^y y + \int [y + \phi(y)]e^y dy + \psi(x - y),$$

or, with a change in the meaning of ϕ,

$$z = x + \phi(y) + e^{-y}\psi(x - y).$$

Monge's Method.

306. The general method of deriving an intermediate equation where one exists is based upon a mode of reasoning similar to the following method for Lagrange's solution of equations of the first order, which is that by which it was originally established.

Given the equation

$$Pp + Qq = R, \quad \ldots \ldots \ldots \quad (1)$$

and the differential relation

$$dz = pdx + qdy, \quad \ldots \ldots \ldots \quad (2)$$

which must exist when z is a function of x and y. Let one of the variables p and q be eliminated, thus

$$Pp + Q\frac{dz - pdx}{dy} = R,$$

or

$$p(Pdy - Qdx) + Qdz - Rdy = 0. \quad \ldots \ldots \quad (3)$$

Hence, the relation between x, y and z which satisfies equation (1) must be such that, when one of the two differential expressions occurring in equation (3) vanishes, the other will in general also vanish. Let us now write the equations

$$\left.\begin{array}{l} Pdy - Qdx = 0 \\ Qdz - Rdy = 0 \end{array}\right\}, \quad \ldots \ldots \quad (4)$$

and suppose $u = a$, $v = b$, to be two integrals of these simultaneous equations. Then $du = 0$ and $dv = 0$ constitute an equivalent differential system, and the relation between x, y and z is such that, if $du = 0$, then $dv = 0$; that is, if u is constant, v is also constant. This condition is satisfied by putting

$$u = \phi(v),$$

which is therefore the solution of equation (1).

Geometrically the reasoning may be stated thus: If upon a surface satisfying equation (1) a point moves in such a way that $Pdy - Qdx = 0$, then also will $Qdz - Rdy = 0$; that is, the point will move in one of the lines determined by equations (4). No restriction is imposed upon the surface, except that it shall pass through these lines, namely, Lagrange's lines defined by $u = a$, $v = b$. The general equation of the surface so restricted is $u = \phi(v)$.

307. Monge applied the same reasoning to the equation

$$Rr + Ss + Tt = V, \quad \ldots \ldots \quad (1)$$

where R, S, T and V are functions of x, y, z, p and q, in connection with which we have, for the total differentials of p and q,

$$dp = rdx + sdy, \quad \ldots \ldots \ldots (2)$$

$$dq = sdx + tdy. \quad \ldots \ldots \ldots (3)$$

Eliminating two of the three variables r, s, t, we have

$$R\frac{dp - sdy}{dx} + Ss + T\frac{dq - sdx}{dy} = V,$$

or

$$Rdpdy + Tdqdx - Vdxdy = s(Rdy^2 - Sdxdy + Tdx^2). \ldots (4)$$

If, then, we can find a relation between x, y, z, p and q, such that, when one of the two differential expressions contained in equation (4) vanishes, the other will vanish also, this relation will satisfy equation (1).

Let us now write the equations

$$\left. \begin{array}{l} Rdy^2 - Sdydx + Tdx^2 = 0 \\ Rdpdy + Tdqdx = Vdxdy \end{array} \right\} \quad \ldots \ldots (5)$$

If $u = a$ and $v = b$ are two integrals of this system, so that $du = 0$, and $dv = 0$ form an equivalent differential system, the required relation will be such that if $du = 0$, then $dv = 0$; that is, if u is constant, v is also constant. As in the preceding article this condition is fulfilled by

$$u = \phi(v),$$

which is now a differential equation of the first order. The integral of this equation is therefore a solution of equation (1).*

---

* The same method applies to the more general form (3), Art. 302, when an intermediate integral exists, but the auxiliary equations are more complex. See Forsyth's Differential Equations, p. 359 *et seq.*

308. The auxiliary equations (5) are known as *Monge's equations*. The first is a quadratic for the ratio $dy:dx$, and is therefore decomposable into two equations of the form $dy = mdx$. Employing either of these the second equation becomes a relation between dp, dq and dx or dy. These two equations, taken in connection with

$$dz = pdx + qdy,$$

form a system of three ordinary differential equations between the five variables x, y, z, p and q. Since four equations are needed to form a determinate system for five variables, it is only when a certain condition is fulfilled that it is possible to obtain by the combination of these three equations an exact equation giving an integral $u = a$. Again, a second condition of integrability[*] must be fulfilled in order that the second integral $v = b$ shall be possible. These two conditions are in fact the same as those mentioned in Art. 303, as necessary to the existence of an intermediate integral containing an arbitrary function.

309. If R, S and T in the given equation contain x and y only, the first of Monge's equations is integrable of itself. Given, for example, the equation

$$xr - (x+y)s + yt = \frac{x+y}{x-y}(p-q). \quad \ldots \quad (1)$$

Monge's equations are

$$xdy^2 + (x+y)dydx + ydx^2 = 0, \quad \ldots \quad (2)$$

$$xdpdy + ydqdx = \frac{x+y}{x-y}(p-q)dydx. \quad \ldots \quad (3)$$

---

[*] When there is a deficiency of one equation in a system, a single condition must be satisfied to make an integral possible, just as a single condition is necessary when one equation is given between three variables. Supposing one integral found, one of the variables can be completely eliminated; there is still a deficiency of one equation in the reduced system, and again a condition must be fulfilled to make a second integral possible.

Equation (2) may be written

$$(dy + dx)(xdy + ydx) = 0.$$

Taking the second factor, we have

$$xdy + ydx = 0,$$

which gives the integral

$$xy = a, \quad \ldots \ldots \ldots (4)$$

and converts equation (3) into

$$\frac{dp - dq}{p - q} = \frac{dx - dy}{x - y}.$$

This gives for the second integral

$$\frac{p - q}{x - y} = b. \quad \ldots \ldots \ldots (5)$$

Hence we have for the intermediate integral

$$\frac{p - q}{x - y} = \phi(xy). \quad \ldots \ldots \ldots (6)$$

To solve this equation of the first order, Lagrange's equations are

$$dx = -dy = \frac{dz}{(x - y)\phi(xy)}, \quad \ldots \ldots (7)$$

of which the first gives

$$x + y = a. \quad \ldots \ldots \ldots (8)$$

For the second integral we readily obtain from equations (7)

$$xdx + ydy = \frac{dz}{\phi(xy)},$$

whence

$$\phi(xy)\,d(xy) = dz.$$

Since ϕ is arbitrary, the integral of the first member is an arbitrary function of xy, hence we may write

$$z - \phi(xy) = \beta; \quad \ldots \ldots \ldots (9)$$

and finally putting $\beta = \psi(a)$, we have

$$z = \phi(xy) + \psi(x+y), \quad \ldots \ldots (10)$$

which is therefore the general integral of equation (1).

Another intermediate integral might have been found, but less readily, by employing the other factor of equation (2).

310. When either of the variables z, p or q is contained in R, S or T, the first of Monge's equations is integrable only in connection with $dz = p\,dx + q\,dy$. For example, given the equation

$$q^2 r - 2pqs + p^2 t = 0.$$

Monge's equations are

$$q^2 dy^2 + 2pq\,dy\,dx + p^2 dx^2 = 0,$$

and

$$q^2 dp\,dy + p^2 dq\,dx = 0.$$

The first is a perfect square and gives only

$$q\,dy + p\,dx = 0,$$

which converts the second into

$$q\,dp - p\,dq = 0.$$

Hence the integrals

$$z = a, \quad \text{and} \quad p = bq,$$

and the intermediate integral

$$p = q\phi(z).$$

For this Lagrange's equations are

$$dx = \frac{-dy}{\phi(z)} = \frac{dz}{0};$$

whence the integrals

$$z = a, \quad \text{and} \quad x\phi(a) + y = \beta;$$

and, putting $\beta = \psi(a)$, the final integral

$$y + x\phi(z) = \psi(z).$$

In this example but one intermediate integral can be found; the form of the final equation is that considered in Art. 301.

311. In the following example, the second of Monge's equations must be combined with $dz = pdx + qdy$. Given

$$r - t = -\frac{4p}{x+y}; \quad \dots \dots \quad (1)$$

for which Monge's equations are

$$dy^2 - dx^2 = 0, \quad \dots \dots \quad (2)$$

$$dpdy - dqdx + \frac{4p}{x+y} dydx = 0. \quad \dots \dots \quad (3)$$

Taking from equation (2)

$$dy - dx = 0,$$

whence the integral

$$y = x + a, \quad \dots \dots \quad (4)$$

equation (3) becomes

$$dp - dq + \frac{4pdx}{2x+a} = 0,$$

or

$$(2x+a)(dp - dq) + 4pdx = 0. \quad \dots \dots \quad (5)$$

To ascertain whether this is an exact equation, subtract from the first member the differential of $(2x+a)(p-q)$, which is

$$(2x+a)(dp - dq) + 2pdx - 2qdx.$$

§ XXIII.] EXAMPLES OF MONGE'S METHOD.

The remainder is
$$2pdx + 2qdx,$$
which, since $dx = dy$, is equivalent to $2dz$. Hence, equation (5) is exact, and gives the integral
$$(2x + a)(p - q) + 2z = b. \quad \ldots \quad (6)$$
From equations (4) and (6) we have the intermediate integral
$$(x + y)(p - q) + 2z = \phi(y - x). \quad \ldots \quad (7)$$
Lagrange's equations now are
$$\frac{dx}{x+y} = -\frac{dy}{x+y} = \frac{dz}{\phi(y-x) - 2z},$$
whence we have the integral
$$x + y = a, \quad \ldots \quad (8)$$
which converts the relation between dy and dz into
$$\frac{dz}{dy} - \frac{2z}{a} = -\frac{\phi(2y - a)}{a}.$$
The integral of this last equation is
$$ze^{-\frac{2y}{a}} = -\frac{1}{a}\int e^{-\frac{2y}{a}} \phi(2y - a)dy + \beta. \quad \ldots \quad (9)$$
Finally using equation (8) and putting $\beta = \frac{1}{a}\psi(a)$, we have
$$(x + y)ze^{-\frac{2y}{x+y}} = -\int e^{-\frac{2y}{a}} \phi(2y - a)dy + \psi(x + y), \quad (10)$$
where $x + y$ is to be put for a after the indicated integration.

312. In this example it was not possible to obtain the second integral required in Lagrange's process in a form containing a simple arbitrary function of the form $\phi(u)$, as was done in finding equation (9), Art. 309. Thus the final integral in the present case is not of the form considered in Art. 300. In the case of a primitive of the present kind, there is but one intermediate integral. Accordingly, it will be found that, had we employed the other factor of equation (2), the resulting system of Monge's equations would not have been integrable.

Examples XXIII.

Solve the following partial differential equations: —

1. $r = f(x, y)$, $\quad z = \iint f(x, y) dx^2 + x\phi(y) + \psi(y)$.

2. $s = \dfrac{x}{y} + a$, $\quad z = \tfrac{1}{2}x^2 \log y + axy + \phi(x) + \psi(y)$.

3. $t - q = e^x + e^y$, $\quad z = y(e^y - e^x) + \phi(x) + e^y\psi(x)$.

4. $p + r = xy$, $\quad z = \tfrac{1}{2}x^2 y - xy + \phi(y) + e^{-x}\psi(y)$.

5. $xr + p = xy$, $\quad z = \tfrac{1}{4}x^2 y + \phi(y)\log x + \psi(y)$.

6. $zr + p^2 = 3xy^2$, $\quad z^2 = x^3 y^2 + x\phi(y) + \psi(y)$.

7. $r + p^2 = y^2$, $\quad z = \log\left[e^{xy}\phi(y) - e^{-xy}\right] + \psi(y)$.

8. $zs - qp = \dfrac{z^2}{xy}$, $\quad z = \phi(x)\psi(y)x^{\log y}$.

9. $s - t = \dfrac{x}{y^2}$, $\quad z = (x + y)\log y + \phi(x) + \psi(x + y)$.

10. $ps - qr = 0$, $\quad x = \phi(z) + \psi(y)$.

11. $x^2 r + 2xys + y^2 t = 0$, $\quad z = x\phi\!\left(\dfrac{y}{x}\right) + \psi\!\left(\dfrac{y}{x}\right)$.

12. $r - a^2 t = 0$, $\quad z = \phi(y + ax) + \psi(y - ax)$.

13. $x^2 r - y^2 t = qy - px$, $\quad z = \phi\!\left(\dfrac{y}{x}\right) + \psi(xy)$.

14. $q(1 + q)r - (p + q + 2pq)s + p(1 + p)t = 0$,
$\quad x = \phi(z) + \psi(x + y + z)$.

15. $(b + cq)^2 r - 2(b + cq)(a + cp)s + (a + cp)^2 t = 0$,
$\quad y + x\phi(ax + by + cz) = \psi(ax + by + cz)$.

XXIV.

Linear Equations.

313. A partial differential equation which is linear with respect to the independent variable z and its derivatives may be written in the symbolic form

$$F(D, D')z = V, \qquad (1)$$

where

$$D = \frac{d}{dx}, \qquad D' = \frac{d}{dy}$$

and V is a function of x and y. We have occasion to consider solutions only in the form

$$z = f(x, y),$$

and shall therefore speak of a value of z which satisfies equation (1) as an *integral*. Since the result of operating with $F(D, D')$ upon the sum of several functions of x and y is obviously the sum of the results of operating upon the functions separately, the sum of a particular integral of equation (1) and the most general integral of

$$F(D, D')z = 0 \qquad (2)$$

will constitute the general integral of equation (1). Hence, as in the case of ordinary differential equations, the general integral of equation (2) is called the *complementary function* for equation (1).

So also, as in the case of ordinary differential equations, when the second member is zero, the product of an integral and an arbitrary constant is also an integral; but this does not, as in the former case, lead to a term of the general integral, since such a term should contain an arbitrary function. It is, in fact, only in special cases that the general integral consists of separate terms involving arbitrary functions.

Homogeneous Equations with Constant Coefficients.

314. The simplest case is that in which the equation is of the form

$$A_0 \frac{d^n z}{dx^n} + A_1 \frac{d^n z}{dx^{n-1} dy} + \ldots + A_n \frac{d^n z}{dy^n} = 0, \quad \ldots \quad (1)$$

the derivatives contained being all of the same order, and their coefficients being constants. Let us assume

$$z = \phi(y + mx).$$

Now $\frac{d}{dx} \psi(y + mx) = m\psi'(y + mx)$ and $\frac{d}{dy} \psi(y + mx) = \psi'(y + mx)$, whatever be the form of the function ψ, therefore the result of substitution, after rejecting the common factor $\phi^{(n)}(y+mx)$, will be

$$A_0 m^n + A_1 m^{n-1} + \ldots + A_n = 0. \quad \ldots \quad (2)$$

Hence, if m be a root of this equation, $z = \phi(y + mx)$ satisfies equation (1), ϕ being an arbitrary function. If $m_1, m_2, \ldots, m_n$ are distinct roots of equation (2), we have the general integral

$$z = \phi_1(y + m_1 x) + \phi_2(y + m_2 x) + \ldots + \phi_n(y + m_n x), \quad (3)$$

where $\phi_1, \phi_2, \ldots, \phi_n$ are arbitrary functions.

Given, for example, the equation

$$\frac{d^2 z}{dx^2} - 3a \frac{d^2 z}{dx\, dy} + 2a^2 \frac{d^2 z}{dy^2} = 0.$$

The equation for m is

$$m^2 - 3am + 2a^2 = 0,$$

whence $m = a$ or $m = 2a$. Hence the general integral is

$$z = \phi(y + ax) + \psi(y + 2ax).$$

315. Equation (1) of the preceding article, when written symbolically, is

$$(A_0 D^n + A_1 D^{n-1} D' + \ldots + A_n D'^n)z = 0,$$

or, resolving into symbolic factors,

$$(D - m_1 D')(D - m_2 D') \ldots (D - m_n D')z = 0. \quad . \quad . \quad (4)$$

Since the factors are commutative, this equation is evidently satisfied by the integrals of the several equations,

$$(D - m_1 D')z = 0, \quad (D - m_2 D')z = 0, \quad \ldots \quad (D - m_n D')z = 0.$$

Accordingly the several terms of the general integral (3) are the integrals of these separate equations.

Again, the equation may be written

$$D'^n f\left(\frac{D}{D'}\right)z = 0, \quad . \quad . \quad . \quad . \quad . \quad . \quad . \quad (5)$$

where f is an algebraic function of the nth degree, and equation (2) is equivalent to

$$f(m) = 0.$$

We may now regard the symbol $\dfrac{D}{D'}$, when operating upon a function of the form $\phi(y + mx)$ as equivalent to the multiplier m, thus

$$\frac{D}{D'}\phi(y + mx) = m\phi(y + mx).$$

It follows that

$$f\left(\frac{D}{D'}\right)\phi(y + mx) = f(m)\phi(y + mx);$$

so that equation (5) is satisfied by $\phi(y + mx)$ when $f(m) = 0$, whatever be the form of the function ϕ.

316. The solution of the component equations, of which the form is

$$(D - mD')z = 0 \quad \ldots \ldots \ldots (1)$$

may be symbolically derived from that of the corresponding case of ordinary differential equations. For, if we regard D' in equation (1) as constant, its integral is

$$z = Ce^{mD'x},$$

where C is a constant of integration. Replacing C by $\phi(y)$, as usual in integrating with respect to one variable only, we have for the symbolic solution

$$z = e^{mxD'}\phi(y), \quad \ldots \ldots \ldots (2)$$

where $\phi(y)$ is written after the symbol because D' operates upon it, though it does not operate upon x. The symbol $e^{mxD'}$ is to be interpreted exactly as if D' were an algebraic quantity. Thus

$$e^{mx\frac{d}{dy}}\phi(y) = \left(1 + mx\frac{d}{dy} + \frac{m^2x^2}{2!}\frac{d^2}{dy^2} + \ldots\right)\phi(y)$$

$$= \phi(y) + mx\phi'(y) + \frac{m^2x^2}{2!}\phi''(y) + \ldots,$$

or

$$e^{mxD'}\phi(y) = \phi(y + mx),$$

by Taylor's theorem, of which this is in fact the symbolic statement (Diff. Calc. Art. 176).

It should be noticed that the process of verifying the identity

$$(D - mD')e^{mxD'}\phi(y) = 0,$$

with the expanded form of the symbol $e^{mxD'}$, is precisely the same as that of verifying

$$(D - m)e^{mx} = 0,$$

with the expanded form of the exponential e^{mx}.

Case of Equal Roots.

317. The general solution, equation (3), Art. 314, contains n arbitrary functions; but when two of the roots of $f(m) = 0$ are equal, say $m_1 = m_2$, the corresponding terms,

$$\phi_1(y + m_1x) + \phi_2(y + m_1x),$$

are equivalent to a single arbitrary function. There is, however, in this case also, a general integral containing n arbitrary functions. To obtain it we need an integral of

$$(D - m_1D')^2 z = 0, \quad \ldots \ldots \ldots (1)$$

in addition to that which also satisfies $(D - m_1D')z = 0$. This required integral will be the solution of

$$(D - m_1D')z = \phi(y + m_1x) ; \quad \ldots \ldots (2)$$

for, if we operate with $D - m_1D'$ on both members of this equation, we obtain equation (1), so that its integral is also an integral of equation (1).

Writing equation (2) in the form

$$p - m_1q = \phi(y + m_1x),$$

Lagrange's equations are

$$dx = -\frac{dy}{m_1} = \frac{dz}{\phi(y + m_1x)},$$

of which the first gives the integral

$$y + m_1 x = a,$$

and then the relation between dx and dz gives

$$z = x\phi(a) + b.$$

Thus the integral of equation (2) is

$$z = x\phi(y + m_1 x) + \psi(y + m_1 x);$$

and, regarding ϕ and ψ as both arbitrary, this is the general integral of equation (1).

318. The solution may also be derived symbolically; for, since the solution of

$$(D - m)^2 z = 0$$

is

$$z = e^{mx}(Ax + B),$$

we have, for the solution of

$$(D - mD')^2 z = 0,$$

$$z = e^{mxD'}[x\phi(y) + \psi(y)],$$

that is,

$$z = x\phi(y + mx) + \psi(y + mx). \quad \ldots \ldots \quad (1)$$

The solution might also have been found in the form

$$z = y\phi_1(y + mx) + \psi_1(y + mx), \quad \ldots \ldots \quad (2)$$

but this is equivalent to the preceding result; for we may write it in the form

$$z = (y + mx - mx)\phi_1(y + mx) + \psi_1(y + mx);$$

and, since $(y + mx)\phi_1(y + mx) + \psi_1(y + mx)$ and $-m\phi_1(y + mx)$

are two independent arbitrary functions of $y + mx$, they may be represented by ψ and ϕ, the equation thus becoming identical with equation (1).

In like manner, if the equation $f\left(\dfrac{D}{D'}\right) = 0$ has r equal roots, the terms corresponding to $(D - mD')^r$ are

$$x^{r-1}\phi_1(y + mx) + x^{r-2}\phi_2(y + mx) + \ldots + \phi_r(y + mx).$$

Case of Imaginary Roots.

319. When the equation has a pair of imaginary roots, $\mu \pm i\nu$, the corresponding terms in the general integral are

$$z = \phi(y + \mu x + i\nu x) + \psi(y + \mu x - i\nu x);$$

or, putting $u = y + \mu x$, $v = \nu x$,

$$\phi(u + iv) + \psi(u - iv).$$

To reduce this expression to a real form, assume

$$\phi_1 = \phi + \psi, \quad \text{and} \quad i\psi_1 = \phi - \psi;$$

so that

$$\phi = \tfrac{1}{2}(\phi_1 + i\psi_1), \quad \text{and} \quad \psi = \tfrac{1}{2}(\phi_1 - i\psi_1).$$

Making the substitutions, the expression becomes

$$z = \tfrac{1}{2}[\phi_1(u + iv) + \phi_1(u - iv)] + \tfrac{i}{2}[\psi_1(u + iv) - \psi_1(u - iv)].$$

In this expression ϕ_1 and ψ_1 are arbitrary functions, since ϕ and ψ were arbitrary; but giving any real forms to ϕ_1 and ψ_1, the two terms are real functions of u and v, that is to say, real functions of x and y.

Given, for example, the equation

$$\frac{d^2z}{dx^2} + \frac{d^2z}{dy^2} = 0,$$

of which the solution in the general form is

$$z = \phi(x+iy) + \psi(x-iy).$$

In the form given above the solution is

$$z = \tfrac{1}{2}[\phi_1(x+iy) + \phi_1(x-iy)] + \frac{i}{2}[\psi_1(x+iy) - \psi_1(x-iy)].$$

If, for instance, we assume $\phi_1(t) = t^3$ and $\psi_1(t) = e^t$, we have the particular solution in real form

$$z = x^3 - 3xy^2 + e^x \sin y,$$

which is readily verified.

The Particular Integral.

320. The methods explained in the preceding articles enable us to find the complementary function for an equation of the form

$$F(D, D')z = V,$$

when $F(D, D')$ is a homogeneous function of D and D', and V a function of x and y. The particular integral, which is denoted by

$$\frac{1}{F(D, D')} V,$$

can also in this case be readily found.

Resolving the homogeneous symbol $F(D, D')$ into factors, we may write

$$F(D, D') = (D - m_1 D')(D - m_2 D') \ldots (D - m_n D'),$$

§ XXIV.]   THE PARTICULAR INTEGRAL.   349

and the inverse symbol $\dfrac{1}{F(D, D')}$ may be separated, as in Art. 105, into partial fractions of the form

$$\frac{N_{\text{1}}}{(D - m_{\text{1}}D')^r},$$

where the numerators are numerical quantities, and r is unity except when multiple roots occur. It is therefore only necessary to interpret the symbol

$$\frac{1}{(D - mD')^r}\Phi(x, y).$$

321. For this purpose we employ the formula

$$\phi(D)e^{ax}V = e^{ax}\phi(D + a)V,$$

proved in Art. 116. Putting mD' in place of a,* this formula gives

$$\frac{1}{D - mD'}\Phi(x, y) = \frac{1}{D - mD'}e^{mxD'}e^{-mxD'}\Phi(x, y)$$

$$= e^{mxD'}\frac{1}{D}\Phi(x, y - mx). \quad \ldots \quad (1)$$

Hence the result is found by subtracting mx from y in the operand, integrating with respect to x, and adding mx to y after the integration. Since

---

* In explanation of this application of the symbolic method, let it be noticed that, just as the formula of Art. 116 is founded upon the equation

$$De^{ax}V = e^{ax}(D + a)V,$$

so the present application of it depends upon

$$De^{mxD'}\Phi(x, y) = e^{mxD'}(D + mD')\Phi(x, y),$$

or

$$D\Phi(x, y + mx) = \text{result of putting } y + mx \text{ for } y \text{ in } (D + mD')\Phi(x, y),$$

which expresses an obvious truth.

$$\int \Phi(x, y - mx) dx = \int^x \Phi(\xi, y - m\xi) d\xi,$$

this may be expressed by the equation

$$\frac{1}{D - mD'} \Phi(x, y) = \int^x \Phi(\xi, y + mx - m\xi) d\xi. \quad . \quad . \quad (2)$$

In like manner, for the terms corresponding to multiple roots of $f(m) = 0$, we have

$$\frac{1}{(D - mD')^r} \Phi(x, y) = \int^x \int \cdots \int \Phi(\xi, y + mx - m\xi) d\xi^r. \quad . \quad (3)$$

322. There are certain methods by which, in the case of special forms of the operand, the result may be obtained more expeditiously than by the general method just given. Some of these, which apply as well when the equation is not homogeneous, will be found in Arts. 328–334. The following applies only when the equation is homogeneous.

Suppose the second member to be of the form $\Phi(ax + by)$. The equation may be written in the form

$$F(D, D')z = D^n f\left(\frac{D'}{D}\right) z = \Phi(ax + by).$$

It is readily seen that

$$f\left(\frac{D'}{D}\right) \Phi(ax + by) = f\left(\frac{b}{a}\right) \Phi(ax + by).$$

We have, therefore, for the particular integral

$$z = \frac{1}{D^n f\left(\frac{D'}{D}\right)} \Phi(ax + by) = \frac{1}{f\left(\frac{b}{a}\right)} \int\!\!\int \cdots \int \Phi(ax + by) dx^n, \quad . \quad (1)$$

or, denoting $ax + by$ by t, since $a^n f\left(\frac{b}{a}\right) = F(a, b)$,

$$z = \frac{1}{F(a, b)} \int\int \ldots \int \Phi(t) dt^n. \quad * \quad \ldots \quad \ldots \quad (2)$$

Given, for example, the equation

$$\frac{d^2z}{dx^2} + \frac{d^2z}{dx\,dy} - 2\frac{d^2z}{dy^2} = \sin(x + 2y),$$

the particular integral is

$$z = \frac{1}{D^2 + DD' - 2D'^2} \sin(x + 2y)$$
$$= \frac{1}{1 + 2 - 8} \int\int \sin t\, dt^2 = \frac{1}{5} \sin t = \frac{1}{5} \sin(x + 2y).$$

Adding the complementary function,

$$z = \phi(y + x) + \psi(y - 2x) + \frac{1}{5} \sin(x + 2y).$$

323. When $F(a, b) = 0$, the operand is of the form of one of the terms of the complementary function. The method then fails, the expression given in the preceding article representing a term included in the complementary function, with an infinite coefficient. In this case, after applying the method to all the factors of the operative symbol, except that which vanishes when we put $D = a$ and $D' = b$, the solution may be completed by means of the formula

$$\frac{1}{D - mD'} f(y + mx) = xf(y + mx),$$

which results immediately from equation (1), Art. 321.

---

* This integral involves an expression of the form $At^{n-1} + Bt^{n-2} + \ldots + L$, in which $A, B, \ldots, L$ are arbitrary constants, but such an expression is included in the complementary function. It must be remembered that the multiple integral in equation (1) is not to be regarded as involving an arbitrary function of y.

Thus, if in the example given in the preceding article the second member had been $f(x+y)$, we should have had

$$z = \frac{1}{D(D-D')} \frac{D}{D+2D'} f(x+y)$$

$$= \frac{1}{3D} \frac{1}{D-D'} e^{xD'} e^{-xD'} f(x+y)$$

$$= \frac{1}{3D} e^{xD'} \frac{1}{D} f(y) = \frac{1}{3D} e^{xD'} x f(y)$$

$$= \frac{1}{3} \int x f(x+y) dx.$$

The Non-Homogeneous Equation.

324. When the equation

$$F(D, D')z = 0 \quad \ldots \ldots \ldots (1)$$

is not homogeneous with respect to D and D', the solution cannot generally be expressed in a form involving arbitrary functions. Let us, however, assume

$$z = ce^{hx+ky}, \quad \ldots \ldots \ldots (2)$$

where c, h and k are constants. Substituting in equation (1), we have, since $De^{hx+ky} = he^{hx+ky}$ and $D'e^{hx+ky} = ke^{hx+ky}$,

$$cF(h, k)e^{hx+ky} = 0.$$

Thus we have a solution of the assumed form, if h and k satisfy the relation

$$F(h, k) = 0, \quad \ldots \ldots \ldots (3)$$

c being arbitrary. Let equation (3) be solved for h in terms of k. Now if $F(h, k)$ is homogeneous, we shall have roots of the form

$$h = m_1 k, \quad h = m_2 k, \quad \ldots, \quad h = m_n k;$$

and, since the sum of any number of terms of the form (2) which satisfy the condition (3) is also a solution, the equation will be satisfied by any expression of the form

$$z = \Sigma c e^{k(y + mx)},$$

where m has any one of the values $m_1, m_2, \ldots, m_n$. But, since for a given value of m this expression is a series of powers of e^{y+mx} with arbitrary coefficients and exponents, it is equivalent to an arbitrary function of e^{y+mx}, that is to say, it denotes an arbitrary function of $y + mx$. This agrees with the result otherwise found in Art. 314.

325. Again, if $F(D, D')$ can be resolved into factors, and one of these is of the form $D - mD' - b$, so that $F(h, k) = 0$ is satisfied by

$$h = mk + b,$$

equation (1) will be satisfied by an expression of the form

$$z = \Sigma c e^{k(y + mx) + bx},$$

where m and b are fixed and c and k are arbitrary. But this expression is equivalent to the product of e^{bx} into an arbitrary function of $y + mx$. Thus, corresponding to every factor of the form $D - mD' - b$ we have a solution of the form

$$z = e^{bx}\phi(y + mx).$$

Given, for example, the equation

$$\frac{d^2z}{dx^2} - \frac{d^2z}{dy^2} + \frac{dz}{dx} + \frac{dz}{dy} = 0,$$

or

$$(D + D')(D - D' + 1)z = 0;$$

the general integral is

$$z = \phi(y - x) + e^{-x}\psi(y + x).$$

We might also have found the solution in the form

$$z = \phi_1(y-x) + e^y\psi_1(y+x);$$

but, writing the last term in the form $e^{y+x-x}\psi_1(y+x)$, this agrees with the previous result if $\psi(t)$ is put for $e^t\psi_1(t)$.

326. In the general case, however, we can only express the solution of

$$F(D, D')z = 0 \quad \ldots \ldots \ldots (1)$$

in the form

$$z = \Sigma c e^{hx+ky}, \quad \ldots \ldots \ldots (2)$$

where

$$F(h, k) = 0, \quad \ldots \ldots \ldots (3)$$

so that c and one of the two quantities h and k admit of an infinite variety of arbitrary values.

Given, for example, the equation

$$\frac{d^2z}{dx^2} - \frac{dz}{dy} = 0.$$

Here $F(D, D') = D^2 - D'$, whence $h^2 - k = 0$, thus the general integral is

$$z = \Sigma c e^{hx + h^2 y}.$$

Putting $h = 1$, $h = 2$, $h = \frac{1}{2}$, etc., we have the particular integrals e^{x+y}, e^{2x+4y}, $e^{\frac{1}{2}x+\frac{1}{4}y}$, etc.

Special Forms of the Integral.

327. There are certain forms of the integral of $F(D, D')z = 0$ which can only be regarded as included in the general expression (2), by supposing two or more of the exponentials to become identical. Let the value of k derived from $F(h, k) = 0$ be

$$k = f(h), \quad \ldots \ldots \ldots (4)$$

then

$$\frac{e^{h_1 x + f(h_1)y} - e^{h_2 x + f(h_2)y}}{h_1 - h_2}$$

is an integral of $F(D, D')z = 0$. When $h_2 = h_1 = h$, this takes the indeterminate form, and its value is

$$\frac{d}{dh}e^{hx+f(h)y} = e^{hx+f(h)y}[x+f'(h)y],$$

which is accordingly an integral. In like manner we can show that $\frac{d^2}{dh^2}e^{hx+f(h)y}$, and in general, $\frac{d^r}{dh^r}e^{hx+f(h)y}$ satisfies equation (1); thus we have the series of integrals

$$\left.\begin{array}{l}ce^{hx+f(h)y}\ [x+f'(h)y]\\ ce^{hx+f(h)y}\{[x+f'(h)y]^2+f''(h)y\}\\ ce^{hx+f(h)y}\{[x+f'(h)y]^3+3f''(h)y[x+f'(h)y]+f'''(h)y\}\\ \ldots\qquad\ldots\qquad\ldots\qquad\ldots\qquad\ldots\qquad\ldots\end{array}\right\}. \quad (5)$$

For example, in the case of the equation $(D^2 - D')z = 0$, the integral e^{hx+h^2y} gives rise to the integrals

$$e^{hx+h^2y}\ (x+2hy),$$
$$e^{hx+h^2y}[(x+2hy)^2+2y],$$
$$e^{hx+h^2y}[(x+2hy)^3+6y(x+2hy)],$$
$$e^{hx+h^2y}[(x+2hy)^4+12y(x+2hy)^2+12y^2],$$
$$\ldots\qquad\ldots\qquad\ldots\qquad\ldots$$

In particular, putting $h = 0$, we have the algebraic integral

$$z = c_1 x + c_2(x^2+2y) + c_3(x^3+6xy)$$
$$+ c_4(x^4+12x^2y+12y^2) + \ldots$$

Special Methods for the Particular Integral.

328. The particular integral of the equation

$$F(D, D')z = V$$

is readily found in the case of certain special forms of the function V.

In the first place, *suppose V to be of the form* e^{ax+by}. Since $De^{ax+by} = ae^{ax+by}$ and $D'e^{ax+by} = be^{ax+by}$, and $F(D, D')$ consists of terms of the form $D^r D'^s$, we have

$$F(D, D')e^{ax+by} = F(a, b)e^{ax+by},$$

or

$$\frac{1}{F(D, D')} F(a, b) e^{ax+by} = e^{ax+by},$$

where $F(a, b)$ is a constant. Hence, except when $F(a, b) = 0$, we have

$$\frac{1}{F(D, D')} e^{ax+by} = \frac{1}{F(a, b)} e^{ax+by}.$$

Thus, when the operand is of the form e^{ax+by}, we may put a for D and b for D', *except when the result introduces an infinite coefficient*. Given, for example, the equation

$$(D^2 - \dot{D}')z = e^{2x+y},$$

the particular integral is

$$z = \frac{1}{D^2 - D'} e^{2x+y} = \tfrac{1}{3} e^{2x+y}.$$

329. In the exceptional case when $F(a, b) = 0$, we may proceed as in Art. 110. Thus, first changing a in the operand to $a + h$, we have

$$z = \frac{1}{F(D, D')} e^{ax+hx+by} = \frac{1}{F(a+h, b)} e^{ax+by} \left(1 + hx + \frac{h^2 x^2}{2} + \ldots \right).$$

The first term of this development is included in the complementary function. Omitting it, we may therefore write for the particular integral

$$z = \frac{h}{F(a+h,\, b)} (x + \tfrac{1}{2}hx^2 + \ldots) e^{ax+by},$$

in which the coefficient takes the indeterminate form when $h = 0$, because $F(a, b) = 0$, and its value is $\dfrac{1}{F_a'(a,\, b)}$, where $F_a'(a, b)$ denotes the derivative of $F(a, b)$ with respect to a. Hence, except when $F_a'(a, b) = 0$, we have

$$z = \frac{x}{F_a'(a,\, b)} e^{ax+by}. \qquad \ldots \ldots \ldots (1)$$

In like manner, if $F_a'(a, b) = 0$, the second term of the development is in the complementary function, and we proceed to the third term. It is evident that we might also have obtained the particular integral when $F(a, b) = 0$ in the form

$$\frac{y}{F_b'(a,\, b)} e^{ax+by}; \qquad \ldots \ldots \ldots (2)$$

but the two results agree, for their difference,

$$\left[\frac{x}{F_a'(a,\, b)} - \frac{y}{F_b'(a,\, b)} \right] e^{ax+by},$$

is readily seen to be included in the first of the special forms (5) of Art. 327, since a and b are admissible values of the h and k of that article.

330. In the next place, *let V be of the form* $\sin(ax + by)$ *or* $\cos(ax + by)$. We may proceed as in Arts. 111 and 112, and it is to be noticed that we have, for these forms of the operand, not only $D^2 = -a^2$ and $D'^2 = -b^2$, but also $DD' = -ab$. Given, for example, the equation

$$\frac{d^2z}{dx^2} + \frac{d^2z}{dx\,dy} + \frac{dz}{dy} - z = \sin(x + 2y),$$

the particular integral is

$$z = \frac{1}{D^2 + DD' + D' - 1} \sin(x + 2y) = \frac{1}{D' - 4} \sin(x + 2y)$$

$$= \frac{D' + 4}{D'^2 - 16} \sin(x + 2y) = -\tfrac{1}{10}[\cos(x + 2y) + 2\sin(x + 2y)].$$

Adding the complementary function, we have

$$z = e^x \phi(y - x) + e^{-x} \psi(y) - \tfrac{1}{10}\cos(x + 2y) - \tfrac{1}{5}\sin(x + 2y).$$

The anomalous case in which an infinite coefficient arises may be treated like the corresponding case in ordinary differential equations.

331. Again *let V be of the form $x^r y^s$*, where r and s are positive integers. In this case, we develop the inverse symbol in ascending powers of D and D'. Thus, if the second member in the example of the preceding article had contained the term $x^2 y$, the corresponding part of the particular integral would have been found as follows:

$$z = -\frac{1}{1 - (D^2 + DD' + D')} x^2 y$$
$$= -[1 + (D^2 + DD' + D') + (D^2 + DD' + D')^2 + \ldots] x^2 y$$
$$= -[1 + D^2 + DD' + D' + 2D^2 D'] x^2 y$$
$$= -x^2 y - 2y - 2x - x^2 - 4.$$

It will be noticed that, on account of the form of the operand, it is unnecessary to retain in the development any terms containing higher powers than D^2 and D'. Again, had the operand been xy, we might have rejected D^2 in the denominator thus:

$$z = \frac{1}{D^2 + DD' + D' - 1} xy = -\frac{1}{1 - D'(1 + D)} xy$$
$$= -[1 + D'(1 + D)] xy = -xy - x - 1.$$

332. When the symbol $F(D, D')$ contains no absolute term, we expand the inverse symbol in ascending powers of either D or D', first dividing the denominator by the term containing the lowest power of the selected symbol. For example, given the equation

$$\frac{d^2z}{dx^2} - 3\frac{d^2z}{dxdy} = x^2y;$$

for the particular integral we have to evaluate

$$z = \frac{1}{D^2 - 3DD'} x^2y.$$

In this case, it is best to develop in ascending powers of D', because, with the given operand, a higher power of D than of D' would have to be retained. Thus

$$z = \frac{1}{D^2\left(1 - 3\dfrac{D'}{D}\right)} x^2y = \frac{1}{D^2}\left(1 + 3\frac{D'}{D}\right)x^2y$$

$$= \frac{1}{D^2}x^2y + \frac{3}{D^3}x^2 = \frac{x^4y}{12} + \frac{x^5}{20}.$$

Adding the complementary function,

$$z = \phi(y) + \psi(y + 3x) + \tfrac{1}{12}x^4y + \tfrac{1}{20}x^5.$$

If we develop the symbol in ascending powers of D, the particular integral found will be

$$z = -\frac{x^3y^2}{18} - \frac{x^2y^3}{54} - \frac{xy^4}{324}.$$

The difference between the two particular integrals will be found to be

$$\tfrac{1}{4860}[(3x + y)^5 - y^5],$$

which is included in the complementary function.

333. Finally, *when the operand is of the form $e^{ax+by}V$*, we may employ the formula of reduction

$$F(D, D')e^{ax+by}V = e^{ax+by}F(D+a, D'+b)V,$$

which is simply a double application of the formula of Art. 116. For example,

$$\frac{1}{D^2 - D'}xe^{ax+a^2y} = e^{ax+a^2y}\frac{1}{D^2 + 2aD - D'}x$$

$$= e^{ax+a^2y}\frac{1}{2aD}\frac{1}{1+\dfrac{D}{2a}}x$$

$$= \frac{e^{ax+a^2y}}{2a}\frac{1}{D}\left(1 - \frac{D}{2a}\right)x = e^{ax+a^2y}\left(\frac{x^2}{4a} - \frac{x}{4a^2}\right).$$

If we develop in powers of D', we shall find

$$\frac{1}{D^2 - D'}xe^{ax+a^2y} = -e^{ax+a^2y}(xy + ay^2).$$

The difference between the two results is accounted for by the special forms given in Art. 327 for the complementary function in this example.

334. As another application of the formula, let us solve the equation

$$\frac{d^2z}{dx^2} + \frac{d^2z}{dxdy} - 6\frac{d^2z}{dy^2} = x^2 \sin(x+y).$$

The particular integral is

$$z = \text{the coefficient of } i \text{ in } \frac{1}{D^2 + DD' - 6D'^2}e^{ix+iy}x^2.$$

Now

$$\frac{1}{D^2 + DD' - 6D'^2}e^{ix+iy}x^2 = e^{ix+iy}\frac{1}{(D+i)^2 + i(D+i) - 6i^2}x^2$$

$$= e^{ix+iy}\frac{1}{D^2 + 3iD + 4}x^2,$$

and by development we find

$$\frac{1}{D^2+3iD+4}x^2 = \frac{x^2}{4} - \frac{3ix}{8} - \frac{13}{32};$$

therefore

$$\frac{1}{D^2+DD'-6D'^2}e^{ix+iy}x^2 = [\cos(x+y)+i\sin(x+y)]\left[\frac{x^2}{4}-\frac{3ix}{8}-\frac{13}{32}\right].$$

Taking the coefficient of i, and adding the complementary function,

$$z = \left(\frac{x^2}{4}-\frac{13}{32}\right)\sin(x+y)-\frac{3x}{8}\cos(x+y)+\phi(y+2x)+\psi(y-3x).$$

Linear Equations with Variable Coefficients.

335. In some cases a linear equation with variable coefficients can be reduced, by a change of the independent variables, to a form in which the coefficients are constant. As an illustration, let us take the equation

$$\frac{1}{x^2}\frac{d^2z}{dx^2} - \frac{1}{x^3}\frac{dz}{dx} = \frac{1}{y^2}\frac{d^2z}{dy^2} - \frac{1}{y^3}\frac{dz}{dy}. \qquad (1)$$

The first member may be written in the form

$$\frac{1}{x}\left[\frac{1}{x}\frac{d^2z}{dx^2} - \frac{1}{x^2}\frac{dz}{dx}\right] = \frac{1}{x}\frac{d}{dx}\cdot\frac{1}{x}\frac{dz}{dx}.$$

Hence, if we put $xdx = d\xi$, whence $\xi = \tfrac{1}{2}x^2$, and in like manner $\eta = \tfrac{1}{2}y^2$, the equation becomes

$$\frac{d^2z}{d\xi^2} = \frac{d^2z}{d\eta^2}. \qquad (2)$$

The integral of this equation is $z = \phi(\xi+\eta)+\psi(\xi-\eta)$; hence that of equation (1) is

$$z = \phi(x^2+y^2)+\psi(x^2-y^2).$$

336. In particular, it is to be noticed that an equation all of whose terms are of the form

$$A x^r y^s \frac{d^{r+s} z}{dx^r dy^s}$$

is reducible to the form with constant coefficients, like the corresponding case in ordinary equations, Art. 123, by the transformations $\xi = \log x$, $\eta = \log y$, which give

$$x \frac{d}{dx} = \frac{d}{d\xi}, \qquad y \frac{d}{dy} = \frac{d}{d\eta}.$$

But, if we put $\vartheta = x \dfrac{d}{dx}$ and $\vartheta' = y \dfrac{d}{dy}$, we may still regard x and y as the independent variables; the transformation is then effected by the formula

$$x^r y^s \frac{d^{r+s}}{dx^r dy^s} = x^r D^r . y^s D'^s = \vartheta(\vartheta-1)\ldots(\vartheta-r+1)\vartheta'(\vartheta'-1)\ldots(\vartheta'-s+1),$$

and the equation reduced to the form

$$F(\vartheta, \vartheta')z = V.$$

The solution of this equation may therefore be derived from that of the equation $F(D, D')z = V$, by replacing x and y by $\log x$ and $\log y$; or it may, as in the following articles, be obtained directly by processes similar to those employed in deriving the solution of $F(D, D')z = V$.

337. Since

$$\vartheta x^r y^s = r x^r y^s, \qquad \vartheta' x^r y^s = s x^r y^s,$$

it is obvious that

$$F(\vartheta, \vartheta') x^r y^s = F(r, s) x^r y^s. \quad \ldots \ldots \quad (1)$$

Hence, if in
$$F(\vartheta, \vartheta')z = 0, \quad \ldots \ldots \ldots (2)$$
we assume
$$z = cx^r y^s,$$
the result is
$$cF(r, s)x^r y^s = 0,$$

and we have a solution of the proposed form if $F(r, s) = 0$. Hence the general solution of equation (2) is
$$z = \Sigma c x^r y^s, \quad \ldots \ldots \ldots (3)$$
where
$$F(r, s) = 0, \quad \ldots \ldots \ldots (4)$$

that is, z is a series in which the coefficients are arbitrary, and the exponents of x and y are connected by the single relation (4).

Now let equation (4) be solved for r in terms of s; if the function $F(\vartheta, \vartheta')$ be homogeneous in ϑ and ϑ', the equation will have roots of the form

$$r = m_1 s, \qquad r = m_2 s, \qquad \text{etc.,}$$

and to each root will correspond a solution of the form

$$y = \Sigma c(yx^m)^s.$$

But this represents an arbitrary function of yx^m. Thus to each factor of $F(\vartheta, \vartheta')$ of the form

$$\vartheta - m\vartheta',$$

there corresponds an independent term of the form

$$z = \phi(yx^m)$$

in the solution of equation (2).

Again, corresponding to a factor of the form

$$\vartheta - m\vartheta' - b,$$

we have the root $r = ms + b$, for $F(r, s) = 0$; and hence the solution $z = \Sigma c(yx^m)^s x^b$, or

$$z = x^b \phi(yx^m).$$

338. For the particular integral of the equation

$$F(\vartheta, \vartheta') = V,$$

we may suppose V to be expanded in products of powers of x and y. By equation (1) of the preceding article, we have

$$\frac{1}{F(\vartheta, \vartheta')} x^a y^b = \frac{1}{F(a, b)} x^a y^b,$$

which gives the particular integral, except when $F(a, b) = 0$. When this is the case, we have, first putting $a + h$ in place of a,

$$\frac{1}{F(\vartheta, \vartheta')} x^{a+h} y^b = \frac{1}{F(a+h, b)} x^a y^b (1 + h \log x + \ldots),$$

or, rejecting the first term of the expansion, which is included in the complementary function, and then putting $h = 0$,

$$\frac{1}{F(\vartheta, \vartheta')} x^a y^b = \frac{1}{F_a'(a, b)} x^a y^b \log x.$$

339. As an illustration, let us take the equation

$$x^2 \frac{d^2z}{dx^2} - y^2 \frac{d^2z}{dy^2} = xy,$$

which, when reduced to the ϑ-form, is

$$\vartheta(\vartheta - 1)z - \vartheta'(\vartheta' - 1)z = xy,$$

or
$$(\vartheta - \vartheta')(\vartheta + \vartheta' - 1)z = xy.$$

The complementary function is
$$\phi(xy) + x\psi\left(\frac{y}{x}\right);$$

for the particular integral,
$$z = \frac{1}{\vartheta - \vartheta'}\frac{1}{\vartheta + \vartheta' - 1}xy = \frac{1}{1+1-1}\frac{1}{\vartheta - \vartheta'}xy$$
$$= \frac{1}{\vartheta - \vartheta'}x^{1+h}y = \frac{1}{h}xy(1 + h\log x + \ldots),$$

or, rejecting the term $\frac{1}{h}xy$ included in $\phi(xy)$, and putting $h = 0$,
$$z = \phi(xy) + x\psi\left(\frac{y}{x}\right) + xy\log x.$$

340. The symbol $\vartheta + \vartheta'$ may be particularly mentioned on account of its relation to the homogeneous function of x and y. Putting
$$\pi = \vartheta + \vartheta',$$
we have $\pi x^r y^s = (r+s)x^r y^s$; hence, if u_n denotes a homogeneous function of x and y of the nth degree, we have
$$\pi u_n = n u_n,$$
where u_n is not necessarily an algebraic function, but may be any function of the form $x^n f\left(\frac{y}{x}\right)$. This is, in fact, the first of Euler's theorem concerning homogeneous functions. See Diff. Calc., Art. 412.

As an example of an equation expressible by means of the single symbol π, let us take

$$x^n \frac{d^n z}{dx^n} + nx^{n-1}y \frac{d^n z}{dx^{n-1}dy} + \frac{n(n-1)}{2} x^{n-2}y^2 \frac{d^n z}{dx^{n-2}dy^2} + \ldots = V. \quad (1)$$

The first member can be shown to be equivalent to

$$\pi(\pi - 1) \ldots (\pi - n + 1)z.$$

Denoting this by $F(\pi)z$, we have

$$F(\pi)u_m = m(m-1) \ldots (m-n+1)u_m, \quad \ldots \quad (2)$$

which, when $F(\pi)$ is expressed as in equation (1), is the general case of Euler's theorem. Thus the complementary function for equation (1) is

$$u_0 + u_1 + u_2 + \ldots + u_{n-1}.$$

Let V contain the given homogeneous function H_m, equation (2) gives for the corresponding term in the particular integral

$$\frac{1}{m(m-1) \ldots (m-n+1)} H_m,$$

except when m is an integer less than n. In this case $F(\pi)$ will contain the factor $\pi - m$, and putting $F(\pi) = (\pi - m)\phi(\pi)$ we readily obtain as in Art. 338

$$\frac{1}{F(\pi)} H_m = \frac{1}{\phi(m)} H_m \log x.$$

EXAMPLES XXIV.

Solve the following partial differential equations : —

1. $2\dfrac{d^2 z}{dx^2} - 3\dfrac{d^2 z}{dxdy} - 2\dfrac{d^2 z}{dy^2} = 0, \qquad z = \phi(2y - x) + \psi(y + 2x).$

2. $\dfrac{d^3 z}{dx^2 dy} - 2\dfrac{d^3 z}{dxdy^2} + \dfrac{d^3 z}{dy^3} = \dfrac{1}{x^2},$

$z = \phi(x) + \psi(x+y) + x\chi(x+y) - y\log x.$

§ XXIV.] EXAMPLES. 367

3. $\dfrac{d^2z}{dx^2} + 5\dfrac{d^2z}{dxdy} + 6\dfrac{d^2z}{dy^2} = \dfrac{1}{y - 2x}$,

$z = \phi(y - 2x) + \psi(y - 3x) + x \log(y - 2x)$.

4. $\dfrac{d^2z}{dx^2} - 3\dfrac{d^2z}{dxdy} + 2\dfrac{d^2z}{dy^2} + \dfrac{dz}{dx} - \dfrac{dz}{dy} = 0$,

$z = \phi(y + x) + e^{-x}\psi(y + 2x)$.

5. $\dfrac{d^2z}{dx^2} - \dfrac{d^2z}{dy^2} - 3\dfrac{dz}{dx} + 3\dfrac{dz}{dy} = e^{x+2y} + xy$,

$z = \phi(x + y) + e^{3y}\psi(y - x)$
$\quad - ye^{x+2y} - (\tfrac{1}{18}x^3 + \tfrac{1}{6}x^2y + \tfrac{1}{6}x^2 + \tfrac{1}{6}xy + \tfrac{2}{27}x)$.

6. $\dfrac{d^2z}{dxdy} + a\dfrac{dz}{dx} + b\dfrac{dz}{dy} + abz = e^{my+nx}$,

$z = e^{-ay}\phi(x) + e^{-bx}\psi(y) + \dfrac{e^{my+nx}}{(m+a)(n+b)}$.

7. $\dfrac{d^2z}{dx^2} - \dfrac{d^2z}{dxdy} + \dfrac{dz}{dy} - z = \cos(x + 2y) + e^y$,

$z = e^x\phi(y) + e^{-x}\psi(x + y) + \tfrac{1}{2}\sin(x + 2y) - xe^y$.

8. $\dfrac{d^2z}{dx^2} - \dfrac{d^2z}{dxdy} + \dfrac{dz}{dx} - \dfrac{dz}{dy} = 2\sin(x + y)$,

$z = e^{-x}\phi(y) + \psi(x + y) + x[\sin(x + y) - \cos(x + y)]$.

9. $\dfrac{dz}{dx} - a\dfrac{dz}{dy} = e^{mx}\cos ny$,

$z = \phi(y + ax) + \dfrac{e^{mx}}{m^2 + n^2a^2}(m \cos ny - na \sin ny)$.

10. $\dfrac{d^2z}{dx^2} - \dfrac{d^2z}{dy^2} + \dfrac{dz}{dx} + 3\dfrac{dz}{dy} - 2z = e^{x-y} - x^2y$,

$z = e^x\phi(y - x) + e^{-2x}\psi(y + x)$
$\quad - \tfrac{1}{4}e^{x-y} + \tfrac{1}{2}x^2y + \tfrac{3}{4}x^2 + \tfrac{1}{2}xy + \tfrac{3}{2}x + \tfrac{3}{4}y + \tfrac{21}{8}$.

11. $mn\dfrac{d^2z}{dx^2} - (m^2+n^2)\dfrac{d^2z}{dxdy} + mn\dfrac{d^2z}{dy^2} + mn^2\dfrac{dz}{dx} - m^2n\dfrac{dz}{dy}$
$$= \cos(kx+ly) + \cos(mx+ny),$$
$$z = \phi(ny+mx) + e^{-nx}\psi(my+nx)$$
$$+ \dfrac{mn\sin(kx+ly) - (mk-nl)\cos(kx+ly)}{(nk-ml)[m^2n^2 + (mk-nl)^2]}$$
$$+ \dfrac{mnx\cos(mx+ny) + (m^2-n^2)x\sin(mx+ny)}{n(m^4 - m^2n^2 + n^4)}.$$

12. $x^2\dfrac{d^2z}{dx^2} + 2xy\dfrac{d^2z}{dxdy} + y^2\dfrac{d^2z}{dy^2} = (x^2+y^2)^{\frac12 n}$,
$$z = \phi\left(\dfrac{y}{x}\right) + x\psi\left(\dfrac{y}{x}\right) + \dfrac{(x^2+y^2)^{\frac12 n}}{n(n-1)}.$$

13. $x^2\dfrac{d^2z}{dx^2} + 2xy\dfrac{d^2z}{dxdy} + y^2\dfrac{d^2z}{dy^2} - nx\dfrac{dz}{dx} - ny\dfrac{dz}{dy} + nz = 0$,
$$z = x^n\phi\left(\dfrac{y}{x}\right) + x\psi\left(\dfrac{y}{x}\right).$$

14. $(x+y)\dfrac{d^2z}{dxdy} - a\dfrac{dz}{dx} = 0$, $z = \int (x+y)^a \phi(x)dx + \psi(y)$.

15. $\dfrac{d^3z}{dx^3} + \dfrac{d^2z}{dy^2} = e^{y-x}(2x - 3y)$,
$$z = \Sigma c e^{a^3y - a^3x} + e^{y-x}(\tfrac13 x^3 + \tfrac23 x - \tfrac34 y^2 + \tfrac94 y).$$

16. Derive the particular integral of
$$\dfrac{d^3z}{dx^3} + \dfrac{d^2z}{dy^2} = e^{y-x}(2x + 3y)$$
in the form $z = xye^{y-x}$.

www.ingramcontent.com/pod-product-compliance
Lightning Source LLC
Chambersburg PA
CBHW030401230426
43664CB00007BB/689